Computer Music Instruments

Victor Lazzarini

Computer Music Instruments

Foundations, Design and Development

 Springer

Victor Lazzarini
Department of Music
Maynooth University
Maynooth
Ireland

ISBN 978-3-319-87573-6   ISBN 978-3-319-63504-0   (eBook)
https://doi.org/10.1007/978-3-319-63504-0

Printed on acid-free paper

This Springer imprint is published by Springer Nature
The registered company is Springer International Publishing AG
The registered company address is: Gewerbestrasse 11, 6330 Cham, Switzerland

*This book is dedicated to John ffitch,
Mathematician, Computer Scientist and
Musician.*

Foreword

I met Victor Lazzarini for the first time in May 2004 in Karlsruhe, during the second International Linux Audio Conference. I was there to present the Faust audio programming language and Victor was giving a talk on developing spectral signal processing applications. What struck me at the outset was Victor's many talents. Victor is not only a highly experienced composer and educator, and Professor of Music and Dean of Arts, Celtic Studies and Philosophy at Maynooth University. He is also a leading researcher in the field of sound synthesis and audio programming languages, and a core developer of the Csound programming language.

It took us some time to transform our highly interesting informal discussions during international conferences into a real collaboration. But in June 2013 Victor suggested we could embed the Faust compiler into the future Csound 6 release in order to be able to interleave Faust and Csound programs. We therefore decided to collaborate to develop a set of opcodes to compile, run and control Faust code directly from Csound (a detailed description of these opcodes can be found in Chapter 2). The resulting Csound 6 opcodes were released in 2014.

Interestingly, it turns out that interleaving programming languages are at the heart of *Computer Music Instruments*. Indeed, the first part of the book introduces the foundations, from audio signals to programming environments, of the three particular programming languages (Python, Csound and Faust) that are used throughout the book.

The second part of the book is dedicated to sound synthesis and signal processing methods. The techniques presented range from source-filter models to frequency-domain techniques, via granular methods, as well as feedback and adaptive systems. Many practical examples with code are provided in Python, Csound and Faust.

The last part of the book focuses on communication protocols, interactions, and computer music platforms. Mobile phones, embedded systems, and Web platforms are covered. The text is completed by two appendices. The first provides support to understand the mathematical notions used in the book. The second provides additional code examples.

I can strongly recommend this comprehensive book. The style is clear and easy to follow. The examples are excellent. Readers of different levels of expertise will gain benefit from it. The great talent of Victor Lazzarini is to provide a book that is interesting from multiple perspectives: computer science, signal processing, software development and, of course, sound art and music!

Lyon, December 2016   *Yann Orlarey*

Preface

If we consider the appearance of Max Mathews' first direct digital synthesis program for the IBM 704 mainframe, MUSIC I, as marking the year zero of computer music, then the field is about to reach its sixtieth anniversary in 2018. Throughout these years, amazing developments in software and hardware technology made the production of (digital) electronic music a common feature of today's music. For the first few decades, access to computers for sound creation was limited to only a few lucky individuals. With the popularisation of microcomputers, a revolution was initiated, allowing a much a wider host of practitioners to avail themselves of these wonderful devices. This has continued today and I expect it will extend well into the future, as new possibilities emerge for music, sound art, sound design, and related activities. The personal computing tools we have at hand today, which have various forms, sizes, and capabilities, are incredibly powerful platforms for the development and manipulation of new sound-making objects. This book is dedicated to the study of these computer instruments from the ground up.

The text is organised into three parts, covering three aspects of the topic: foundations; instrument development from the perspective of signal processing; and the design of applications on existing computer music platforms. The first chapter, in Part I, explores basic concepts relating to audio signals in and out of the computer. It provides a gentle introduction to some key principles that will be used throughout the book, with a touch of mathematics and plenty of illustrative examples. This is followed by an introduction to the programming tools that will be used in the following two parts of the book, Python, Csound, and Faust. These three languages were judiciously chosen to cover a wide variety of applications, which range from demonstrating signal-processing algorithms to implementing full applications. Another important feature of these three environments is that they can be nicely interleaved, from a higher to a lower level, and from an application-development to a signal-processing function. This is demonstrated by a full example at the end of the chapter.

The second part of the book is fully devoted to exploring ways of making sound with a computer, the signal-processing design for instruments. It includes some classic approaches, such as source-filter models, clever ways of manipulating mathematical formulae, newer approaches to feedback and adaptive techniques, and the methods of granular synthesis. Part II is complemented by a thorough examination of the principles of frequency-domain analysis-synthesis principles. Although these chapters tend to make continued use of mathematical expressions, the discussion also employs graphical plots and programming examples to complete the exploration of each technique. This three-pronged approach is aimed at providing a full perspective and a thorough examination of these sound-generation algorithms.

Finally, in the third part, we look at some complementary aspects of instrument design: interaction tools and development platforms. Chapter 8 explores various means of communication with sound synthesis programs, with examples presented in the three target programming environments. We also examine graphical user interfaces and the principles of custom hardware for sound control. The final chapter in the book then discusses the various platforms for the development and performance of computer instruments.

This book is designed for readers of different levels of expertise, from musicians to sound artists, researchers, software developers, and computer scientists. More experienced users may want to skip some of the introductory points in Part I, depending on their background. Readers who are unsure about the mathematical language used can avail themselves of Appendix A where all concepts applied in the book are thoroughly explained (assuming just a basic understanding of arithmetics). All important code that is not completely provided in the main text of the book appears fully in Appendix B and is referenced in the relevant chapters.

I hope that this book will prove to be a useful reference for computer music practitioners on the topic of instrument development. It aims to provide a solid foundation for further research, development, music-making, and sound design. Readers are encouraged to experiment with the examples and programs presented here as they develop new perspectives in their practice of sound and music computing.

Maynooth, December 2016   *Victor Lazzarini*

Acknowledgements

I would like to acknowledge the help and encouragement of many of my colleagues at the university and in the computer music community. In particular, I would like to thank some of my close allies in the Csound brotherhood, Steven Yi, Øyvind Brandtsegg, Joachim Heintz, Tarmo Johannes, Rory Walsh, Iain McCurdy, Richard Boulanger and Michael Gogins, from whom I have learned a lot throughout these many years of interaction. In addition, I wish to acknowledge the work of François Pinot in designing such a great Python interface to the Csound API, ctcsound, a much better fit to the system than the automatically-generated wrappers we used to have.

I would also like to mention my colleagues Joe Timoney and Tom Lysaght, at the Computer Science Department, and Rudi Villing, at the Electronic Engineering Department, in Maynooth, to whom I am very thankful for their collaboration and support. Likewise, it is important to mention my co-researchers in the Ubiquitous Music (ubimus) group, Damián Keller, Nuno Otero and others, with whom I have had numerous exchanges that cemented many ideas underpinning my work in this field.

Some of the original research that led to parts of this book was also conducted in collaboration with colleagues from Aalto University in Finland, and I would like to express my gratitude to Jari Kleimola, Jussi Pekkonen, and Vesa Välimäki for their part in this work. Thanks should also go to the Faust development team at GRAME, led by Stéphane Letz and Yann Orlarey (who very kindly contributed a foreword to this book). Theirs is a fantastic piece of work, providing such a wonderful software system for computer music.

This book would not exist without the support, help, and understanding of my family, who many times had to endure my (mental, if not physical) disappearance for hours at weekends and evenings, as I battled with typesetting, programming, and explaining my ideas more clearly. To them, Alice, my wife, and our children Danny, Ellie, and Chris, I would like to make a special dedication.

I am very thankful to my editor at Springer, Ronan Nugent, for facilitating the development of this and other projects, and for the very efficient manner in which these have been brought to fruition. The peace of mind provided by a well-structured process allows for a much more enjoyable writing experience.

Finally, I would like to pay tribute to John ffitch, to whom this book is dedicated. John has been a very influential figure in the computer music community and has given us a huge contribution in terms of free software, expertise, and guidance. This is just a small token of our gratitude for his generosity and support.

Contents

Part III Application Development

Appendices

Acronyms

0dbfs	Zero decibel full scale
12TET	12-Tone Equal Temperament
ADC	Analogue-to-Digital Converter
ADSR	Attack-Decay-Sustain-Release
AM	Amplitude Modulation
AP	All Pass
API	Application Programming Interface
BP	Band Pass
bpm	beats per minute
BR	Band Reject
CDP	Composer's Desktop Project
CLI	Command-Line Interface
cps	cycles per second
CPU	Central Processing Units
DAC	Digital-to-Analogue Converter
DAW	Digital Audio Workstation
dB	Decibel
DFT	Discrete Fourier Transform
DMI	Digital Music Instrument
DSL	Domain Specific Language
DSP	Digital Signal Processing
FBAM	Feedback Amplitude Modulation
FFT	Fast Fourier Transform
FIR	Finite Impulse Response
FIFO	First In First Out
FM	Frequency Modulation
FOF	Fonction d'Onde Formantique, formant wave function
FT	Fourier Transform
GPGPU	General Purpose Graphic Programming Unit
GPIO	General-Purpose Input/Output
GPPL	General-Purpose Programming Language

GPU	Graphic Programming Unit
GUI	Graphical User Interface
HDMI	High Definition Monitor Interface
HP	High Pass
HTML	Hypertext Markup Language
HTTP	Hypertext Transfer Protocol
Hz	Hertz
IDE	Integrated Development Environment
IDFT	Inverse Discrete Fourier Transform
IF	Instantaneous Frequency
IIR	Infinite Impulse Response
IO	Input-Output
IP	Internet Protocol
IR	Impulse Response
ISTFT	Inverse Short-Time Fourier Transform
JIT	Just In Time
JS	Javascript
LADSPA	Linux Audio Simple Plugin Architecture
LP	Low Pass
LTI	Linear Time-Invariant
MIDI	Musical Instrument Digital Interface
ModFM	Modified Frequency Modulation
MTU	Maximum Transmission Unit
OS	Operating Systeml
OSC	Open Sound Control
PAF	Phase-Aligned Formant
PD	Phase Distortion
PLTV	Periodic Linear Time-Varying
PM	Phase Modulation
PNaCl	Portable Native Client
PV	Phase Vocoder
PWM	Pulse Wave Modulation
RMS	Root Mean Square
RTP	Realtime Transport Protocol
SpSB	Split-Sideband
STFT	Short-Time Fourier Transform
TCP	Transmission Control Protocol
UDO	User-Defined Opcode
UDP	User Datagram Protocol
UG	Unit Generator
UI	User Interface
VDU	Video Display Unit
VGA	Video Graphics Array
VST	Virtual Studio Technology
WIMP	Windows, Icons, Menus, Pointer

Part I
Foundations

Chapter 1
Audio and Music Signals

Abstract The fundamental aspects of audio and musical signals are discussed in this chapter, as a preamble to exploring how computer instruments can be created. A mostly non-mathematical, very intuitive notion of signals, waves, spectra and their principal characteristics is first developed, introducing the reader to these concepts. This is followed by a clear delineation of the case of musical signals, with their specific conditions. A more formal look at signals is then attempted, looking at these from the perspective of computer-based representations.

We use the general term *signals* to designate a number of diverse things that can be described mathematically by functions of one or more independent variables [40, 76]. Functions are taken here in a wide sense as constructs that map one or more parameters to a given output. For instance, we can think of the act of turning a knob from maximum to zero over a certain amount of time as a function that maps time to changes in the knob position (Fig. 1.1). The signal, in this case, is something that causes, or controls, these changes.

In the specific case of computer instruments, we can narrow this definition down a little more. Signals are the information-carrying inputs and outputs of processes in a system [72]. Such a system can be limited to software running on a given hardware, or it can incorporate external components (and processes) such as our hearing mechanism, a sound propagation medium (e.g. the air), transducers (e.g. loudspeakers and microphones), and mechanical devices (e.g. a guitar).

In this book, we will be concerned mostly with signals in the software/hardware environment of a computing platform. However, the fact that in many cases these are representations, models or encoded forms of externally-occurring signals is important and for this reason we should begin by looking at some key concepts that will apply to audio and musical signals more generally. The specific details of how mechanical systems such as the air medium and our hearing mechanism work are, nevertheless, beyond the scope of this text. We will be interested mostly in how we can describe, emulate and manipulate signals that are eventually decoded and transduced into a form that can be taken in by our auditory system.

© Springer International Publishing AG 2017
V. Lazzarini, *Computer Music Instruments*,
https://doi.org/10.1007/978-3-319-63504-0_1

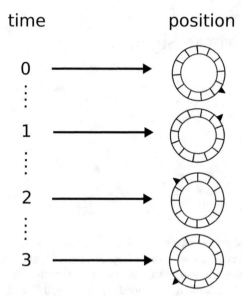

Fig. 1.1 A function mapping time to position. This could be used to describe a signal that controls the position of a knob over a certain time duration.

1.1 Audio Signals

Signals as defined above can take many forms, even if we consider them only in the narrower sense of an information-carrying link between two objects The forms we are interested in are the ones that can eventually be made into a mechanical disturbance of a medium such as the air and have a number of important characteristics that allow them to be picked up by our ears. This allows us to home in on a much more constrained definition of something we will call an *audio* signal.

Let's start by delimiting the desired characteristics of the final form of these signals. Firstly, when we say *mechanical* disturbances, we imply that something is put into motion. In this case, we are talking about the matter that composes the medium in which the signals propagate from source to destination. The most common medium is the air and, even though there are many alternatives to it, this is the one that ultimately matters in this particular discussion as our ears are designed to operate in it. To sum up, we are talking about signals that are carried by the movement of matter that makes up the air medium (Fig. 1.2) [5, 31, 9].

Fig. 1.2 A representation of particles in the air and their movement carrying an audio signal.

1.1.1 Waves

We now come to an important abstraction that helps us explore and understand signals that are travelling in a medium. From a general perspective we can idealise the movement of the medium as *waves* that are traversing it [107, 5], starting from a source and then reaching a point where they can be picked up. It is important not to lose sight of the fact that this is an abstraction that is constructed by connecting all the individual movements of the components of the medium (taken to be, in a generic sense, its molecules or particles, which is yet another abstraction), as illustrated in Fig. 1.3.

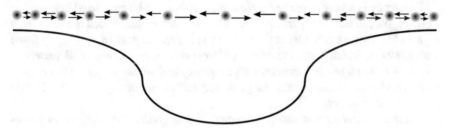

Fig. 1.3 The wave as an abstraction of the movement of particles in the air carrying an audio signal. In this particular case, we are tracking the compressions and decompressions that occur at different points in space as the signal travels through the medium.

So waves are a compact description of how air is, in turn, compressed and rarefied as the medium is put into motion [106]. Or, from a complementary perspective, how particles are made to speed up, slow down, stop, change direction, speed up, slow down, etc., back and forth. The first view is that of a pressure wave that propagates through the medium, whereas the second is that of a velocity wave [31]. They describe the same phenomenon, but provide two different ways of categorising it. Both of them relate to things that happen over time and so, going back to our first definition of signals, such waves can be described by functions of time.

This abstraction is very useful not only because it captures the movement very well, but also because it leads to ways of analysing the characteristics of the signal that is travelling through the medium. It is so powerful that we can make the identity signal = wave without thinking too much about it, even though they are not strictly the same thing. This goes even further, as we will be able to extend this idea to our computer models of a signal, also calling them waves, or *waveforms*.

1.1.2 Parameters

The basic characteristics that define an audio waveform, and by extension an audio signal, can be laid down as parameters of one or more functions that describe or at least approximate it. Depending on the complexity of the wave in question, we may be able to model it with a varying number of these parameters and functions. Starting with simple signals, one of the things we observe is that they can be *periodic*: the back-and-forth movement captured by the concept of a wave may repeat over time [8].

In this case, the amount of time that it takes to complete the repeated pattern (called a wave cycle) is an important parameter that can describe this signal very well. This is called the *period* of the waveform (Fig. 1.4), and we can ask a complementary question about it: how many of these cycles will occur in a given duration (say over one second)? This is actually the reciprocal of our first parameter, and is called the *frequency* of the waveform.

Frequency is a very important parameter, one that will be abstracted in a number of ways. We can intuitively link it to our sense of pitch height, which is a scale from high to low that we often use to classify sounds that we hear (not to be confused with loudness, which is connected to a different parameter). In a general sense, our sensation of pitch height increases with frequency and we are only able to perceive signals that exist within a certain range of frequencies, nominally from 20 to 20,000 cycles per second (cps[1]) [31].

Another parameter that can be identified in waveforms is the amount of movement in the air particles. This can describe how wide the range of compression/decompression that takes place is, or, from the velocity perspective, how fast the particles are moving. This is linked to how much energy the source is pushing into the medium, and we perceive it as the amount of loudness or the volume of the signal. This parameter can be measured either as the maxima of the peaks of compression/decompression (or velocity), called the *peak amplitudes* (Fig. 1.4), or as an average over time, the *RMS*[2] *amplitude*. We are able to perceive tiny amounts of pressure variation, but this depends on the frequencies of the waveform.

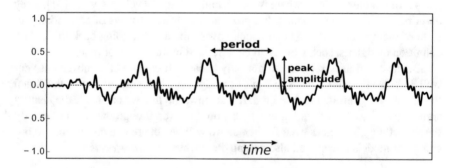

Fig. 1.4 A plot of a waveform showing its period and peak amplitude.

---

[1] frequency is also measured in Hertz (Hz); 1 Hz is equivalent to 1 cps

[2] RMS stands for root-mean-square, which is a method of calculating the average of a varying quantity, such as a waveform [68].

These two parameters are fundamental to all types of waveforms, but there are others which we will explore as we introduce a more detailed description of audio signals. Also, we have not explored signals that have no periodicity, or whose periods are too long for our ears to integrate as a single waveform cycle. For these cases, we will need to generalise the concept of frequency that we have introduced here. This will allow us also to start considering another dimension of audio signals, which has to do with characteristics that are to do with neither pitch height nor amplitude/loudness. We often use the word *colour* to denote this perceptual aspect of audio waves, as we have no alternative but to look for metaphors to describe what we hear.

1.1.3 Waveform shapes

We can abstract even further in order to contemplate how different colours can arise out of an audio waveform. This time, we look at the shapes that the particle movements create if we were to draw up ('plot') their changes in pressure or velocity. Suppose we look at a given point in space and plot the compression and rarefaction that is taking place over time as a wave travels through it. Regardless of the periodicity of the signal (or lack thereof), we will see the emergence of a shape that defines the waveform. If there is a certain amount of repetition over time, that is, a certain general pattern appears to exist, then we are likely to be able to hear a pitch in this signal. We will also be able to see its peak amplitude, which will be determined by the height of the plot (e.g. the maximum compression/decompression at that point).

Once we have abstracted the waveform (and signal) in this form, we do not need to keep going back to its original physical form. This is very convenient because it will be this model that we will be working with inside the computer and this will be a valid representation (within some limits, as we will see) of an audio signal that can be apprehended by our ears. We will come back to these ideas later on.

More generally, the shape of the waveform is linked to its colour characteristics. If the periods are constant and the peak amplitude is also unchanging, but the shape of the repeated patterns varies somewhat, we will perceive a modification in the quality of the sound that is this third, more volatile, percept called *timbre* [31, 106]. This is more of an emergent and multidimensional [29] quality than an easily-definable parameter of a signal, but it is a very important aspect of the signal, especially when we come to consider musical signals.

Determining the timbre of a waveform is a complex task. However it is possible to develop a certain number of concepts that can help us to determine how different shapes arise. From these, we can abstract (again!) a number of principles that can allow us to understand various aspects of the phenomena of timbre, even if some of these might be more elusive than others.

1.1.4 Breaking waveforms down

To start exploring the various shapes of audio signals, and how these can relate to our perception of timbre, we need to look first at very simple waves. The advantage of these is that they can be composed to create more complex forms, and thus allow us to generalise the notions we have developed so far. The ideal form that these simple signals take is defined by well-known mathematical functions called *sinusoids*. These are fundamental to describing a variety of physical phenomena and will be very useful here.

It is possible to derive a sinusoid by plotting circular motion, for instance the height assumed by a spoke in a wheel as it rotates (Fig. 1.5) [3, 22]. The parameters of a waveform introduced in Section 1.1.3 are featured here: the speed of the rotation will determine the frequency and the width of the wheel is linked to the amplitude. If we wished, we could use this rotating wheel to produce a 'real' audio signal directly by making it induce a current to drive a loudspeaker[3]. However, it will be better to proceed by using abstractions, which are more useful for making general observations about waves and signals.

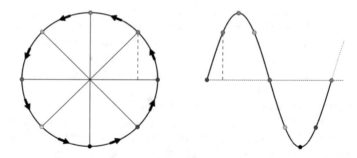

Fig. 1.5 Plotting the vertical position of a spoke in a wheel as it rotates yields a sinusoidal waveform.

In addition to amplitude and frequency, there is a third parameter associated with sinusoids, called their *phase*. This measures the amount of synchronisation between two sinusoids of the same frequency, by defining the starting point of the waveform cycle. If it starts from zero and rises, then we say the wave is in *sine* phase. If it starts from the maximum peak amplitude and falls, then it is in *cosine* phase. The actual phase of a sinusoid can be anything in between these two (including inverted sine and inverted cosine phases).

The reason that these two particular starting positions (cosine and sine) are so important is that they are at right angles to each other. If we plot the various vertical

---

[3] This is indeed a method used by some electric instruments, such as the Hammond organ.

positions that a spoke in a wheel can assume as it rotates, we can also plot the various horizontal positions that it assumes. When we look at these in relation to each other, the former will show the wave in sine phase, and the latter in cosine phase, as shown in Fig. 1.6. Or, more compactly, we can say that the first is a sine wave and the second a cosine wave. Following from this right-angle 'sync' is an important fact: we can describe any sinusoid as a weighted sum of a sine and a cosine wave (of the same frequency, of course) [98, 68].

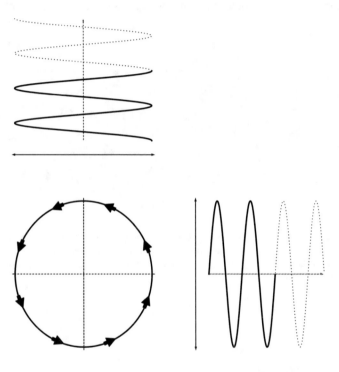

Fig. 1.6 Two phases of a sinusoid: plotting the vertical position yields a sine wave, whereas if we take the horizontal position, we get a cosine wave.

So now we have a simple signal, the sinusoid, which is characterised by its frequency, amplitude and phase. The next abstraction is to think of this as a kind of building block for more complex waveforms. In fact, we can go even further and, in general terms, consider that any audio signal can theoretically be broken down into sinusoidal components. There might be many of them, some of them might be of very short duration, and this decomposition might be very complicated, but the principle still applies.

This concept then allows us to think of an audio signal as being composed of a mix of sinusoids at different frequencies, each one with a given amplitude and phase. Often it is the case that we go one step further in this abstraction and remove the mention of sinusoids, using the term *frequencies* directly: for example, we can say that a signal has significant energy in certain frequencies, or discuss the low frequencies that compose another signal, etc. The set of components that make up a complex waveform is called its frequency *spectrum*. A useful representation of these frequencies that a signal is made up of is seen in a spectrogram, where the distribution of energy (amplitudes) in different parts of the spectrum is plotted over time (Fig.1.7).

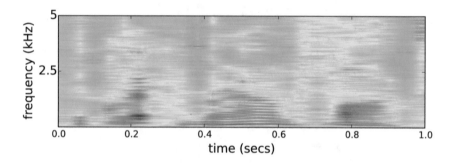

Fig. 1.7 A spectrogram showing the distribution of energy across frequencies in the spectrum. Darker shades of red indicate frequencies with higher amplitudes, whereas lighter yellow/-green/blue areas show low-energy components or their absence. Frequencies range from 0 to 5,000 cycles per second (or Hz).

1.2 Musical Signals

If we think of audio signals as a specific category of signals in general, which have the attributes discussed in Section 1.1, then *musical* signals are a subcategory of these. However, their specific characteristics do not make them a smaller subset of all audio signals. In fact, what we could describe as *musical* includes a quite extensive range of objects. The important distinction is that while an audio signal in general might not require a certain fidelity in transmission, coding/decoding, storage, and playback, musical signals often do.

Let's start with an example. In order for us to understand what is communicated in a signal that carries speech information, we do not need a very wide range of frequencies. The low end does not need to fall below 100 Hz, and the highest frequencies can be accommodated below 4,000 Hz (but speech can be intelligible within an even narrower range). Likewise, we do not need to have a particularly large range of amplitudes, as the information can be conveyed without much change in volume.

It is not possible, however, to limit musical signals to these constraints. Of course, there might be some signals that could well fit them, but in general, music requires a fuller range of frequencies and amplitudes. For this reason, devices and equipment that are designed for musical applications need to be able to operate with greater precision and quality. It is often the case, for instance, that ordinary telephone lines are not good enough for the transmission of music, as they are specifically designed for speech.

1.2.1 Scales

In order to manipulate musical signals, it is useful to find a means of organising some of their basic attributes, such as amplitude and pitch. For this purpose, we can make use of well-defined standard *scales*. Both pitch and amplitude are perceived logarithmically, that is, on the basis of ratios of values rather than their differences (see Appendix A, Section A.1). For instance, we hear a given change in amplitude, say from 0.25 to 0.5, as a similar step to that from 0.1 to 0.2, a doubling of volume. Similarly, a change from 100 to 200 Hz is perceived as similar to 250 to 500 Hz (an interval of an octave).

To facilitate measurements, we can build scales that are based on linear units, defining differences in amplitudes (rather than ratios). For this we need to define them in terms of a logarithm. So, for amplitudes we have the decibel (dB) unit, based on the following expression:

$$20\log_{10}\frac{A_1}{A_2} \tag{1.1}$$

This yields the difference of the amplitudes A_1 and A_2. A 6 dB change means a doubling (positive difference) or halving (negative difference) of amplitudes. The dB expression gives the units and the means to build a scale. To complete this, we need to choose a reference value (e.g, the A_2 in Eq. 1.1). In digital audio, an obvious choice is full scale amplitude, which is the maximum numeric value allowed (see Section 1.3.1). This sets the 0 dB value for the scale and all the other values are therefore in the negative range. Half-scale amplitude is then set to −6 dB, quarter-scale to −12 dB, etc. A simple case is when 0 dB is set to 1, and the expression defining the scale becomes

$$A_{dB}(x) = 20\log_{10}x \tag{1.2}$$

The amplitude scale based on dB units and defined with reference to a maximum value is called the *0dBFS* scale (Fig. 1.8).

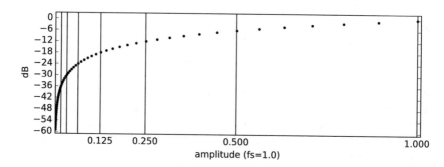

Fig. 1.8 A plot of amplitudes on the 0dBFS scale, with a full-scale reference of 1.0, from -60 to 0 dBFS.

Likewise, pitch scales can also be defined in logarithmic terms. The most common unit is based on a 12-step division of the octave, which is the pitch ratio 2:1. In this system, octave-separated pitches belong to the same class (pitch class) and share their octave-invariant name. Based on this, an interval between two pitches P_1 and P_2 (given in cps, or Hz) is measured by

$$12 \log_2 \frac{P_1}{P_2} \tag{1.3}$$

These units can be used to build the 12-tone equal-tempered (12TET) scale, with twelve pitch classes. Seven of these are given letter names A-G, and the five remaining are denominated by using the nearest pitch class letter, adding a sharp ($\sharp$) in front, if the pitch is one step above it or a flat ($\flat$), if it is one step below (therefore these pitch classes can be named using letter names that are either side of the letter name, see Table 1.1). The standard reference point for the 12TET scale is the pitch A in octave 4 (A4). This can be set to a given frequency and the frequencies $f(n)$ for other pitches are found using an expression based on Eq. 1.3:

$$f(n) = f_{A4} \times 2^{\frac{n}{12}} \tag{1.4}$$

where n is the number of 12-tone steps between a given pitch (in 12TET) and A4 (Fig. 1.9).

Table 1.1 Pitch-class names and respective frequencies on the 12TET scale: one octave from A3 to A4.

Pitch class	A(3)	A$\sharp$ B$\flat$	B C$\flat$	C B$\sharp$	C$\sharp$ D$\flat$	D	D$\sharp$ E$\flat$	E F$\flat$	F E$\sharp$	F$\sharp$ G$\flat$	G	G$\sharp$ A$\flat$	A(4)
Frequency (Hz)	220	233	246.9	261.6	277.2	293.7	311.1	329.6	349.2	370	392	415.3	440

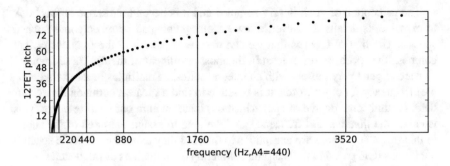

Fig. 1.9 A plot of pitches on the 12TET scale, with A4 reference at 440 Hz, from A0 to C8.

A finer-grained unit ς for an interval ratio p is obtained by subdividing each one of the 12 steps into 100 parts,

$$\varsigma = 1200 \log_2 p \tag{1.5}$$

This unit is called a *cent* and can be used to define small differences in pitches. Frequency scales are the object of study in the field of tuning and temperament. A variety of approaches can be used to construct them, which can produce scales with different numbers of steps and step sizes [9].

1.2.2 Spectral types

In terms of their general characteristics, musical signals are not bound to specific spectral descriptions (as in the case for instance, of speech). The class of objects that compose them is limited only by our hearing capacity. There are however some specific spectral types[4] that are worth discussing in more detail. The first one of these, which is very common in many musical practices, is that of signals that have a *pitch* characteristic, or more precisely, that evoke the sense of pitch height.

Definitions are fairly fluid here, as we are dealing with a perceptual continuum. However, what we will call here *pitched spectra*, are sounds that lead to the possibility of ascribing a specific pitch height to them that is unambiguous. The most extreme case of this is of a pure sinusoid, which is a single frequency (rather than a mixture of them). More commonly, we will find that this spectral type is composed of several frequencies, which, when combined, will lead us to assign a specific pitch to it.

---

[4] We are using this terminology in a loose way, to describe signals that are alike in some recognisable form. The term 'spectral typology' has been used elsewhere in a more precise manner in the analysis of musical works, but this is not the intended meaning here [96].

The perception of pitch is very complex and it is beyond the scope of this book to examine its details. However, a few general points can be made about pitched spectra. The first of them is that we can observe a pattern in the frequencies that compose the spectrum (or at least in the most prominent of them). These frequencies are aligned very closely with an integral series, as multiples of a certain (audible) frequency. In other words, it is possible to find a greatest common divisor for them (within some degree of approximation) that is within our hearing range. This is called the *fundamental* frequency, and we tend to assign the pitch of the signal to this frequency. There are exceptions to this, but they depend on specific spectral conditions (e.g. gaps in the spectrum, conflicting combinations of frequencies, etc.). A plot of a typical harmonic spectrum is shown in Fig. 1.10, where frequencies are clearly displayed with an equal spacing, corresponding to multiples of a fundamental (261 Hz in this case). The fluctuation seen in the components is due to vibrato in the performance.

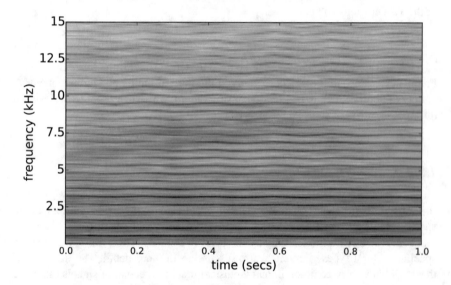

Fig. 1.10 A plot of the pitched spectrum of a 1-second viola sound, where each frequency component is clearly seen as a dark line on the spectrogram.

This type of spectral alignment can also be called a *harmonic* spectrum. This comes from the name given to component frequencies that are integer multiples of a fundamental, *harmonics*. A set of frequencies that are multiples of a fundamental frequency is called a *harmonic series*, and we can say that frequencies in such a spectrum are harmonically related. It is also important to note that these conditions will also give rise to a periodic waveform, the period of which is the reciprocal to the fundamental frequency. So a periodic waveform with a fundamental in the correct range will evoke the sense of a given pitch height. In fact, by repeating a pattern at a fast enough rate, it is possible to create a pitched spectrum, even if the original played more slowly in one single shot does not evoke a pitch.

There are a variety of forms that pitched spectra can take, the simplest being of a straight, static over time, harmonic series (Fig. 1.11). In some other cases, which are commonly found in real-life sounds, spectra are more dynamic, with frequencies appearing and disappearing as time passes. In fact, the fundamental frequency might not actually be present as a component in the spectrum, but might be evoked by the combination of harmonics. In some situations, the frequencies in a pitched spectrum can be slightly misaligned with respect to the exact harmonic series, but the resulting sound is still pitched. A typical example of this is found in a piano tone, whose spectrum is stretched compared to a normal harmonic series.

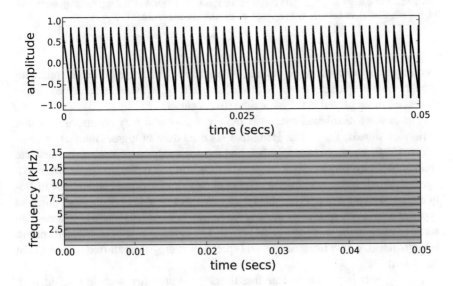

Fig. 1.11 A plot of a sawtooth wave and its spectrum, showing the relationship between a harmonic spectrum and a periodic waveform.

In other cases, we may find spectra that are quasi-pitched. In other words, there is increased ambiguity in assigning a pitch to the sound. This can happen when there is a mix of components that are closely aligned with a harmonic series alongside frequencies that are not. Also, in cases where there are only a few components, not harmonically related, with a dominant frequency among them. As discussed before, there is a continuum between pitched and quasi-pitched spectra, rather than a definite boundary between these types.

Equally, it is also possible to speak of a spectral type that we might call *inharmonic*. This is the case for signals whose frequencies are not in a harmonic relationship. Among these, we find two distinct cases: spectra with discrete, clearly separated components, and *distributed* spectra. The former case is a continuous ex-

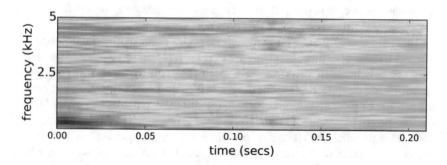

Fig. 1.12 A spectrogram of two cymbal strikes at 0 and 0.11 secs. We can see how there is a distribution of energy across certain bands of the spectrum (with dark red representing areas with higher amplitude), rather than the presence of individual components.

tension of the quasi-harmonic sounds, where there is no sense of definite pitch, although some frequencies might be more dominant. Often these are perceived almost as several tones played at the same time (like a *chord*).

The case of distributed spectra is more complex and may encompass a wider variety of signals (Fig. 1.12). Here, there is a mixture of a great number of components that are very close together, and cannot easily be separated into discrete frequencies, covering a certain range of the spectrum. A typical example is that of a cymbal sound, whose spectrogram shows energy dominating a certain band of frequencies, which gives it its audible quality. We will be able to distinguish different types of distributed spectra by the way the energy is spread, but there is no emerging sense of pitch in the ways discussed before. The extreme case of distributed spectra is *white noise*, which features an equal spread of energy over all frequencies (a flat spectrum).

Finally, it is important to note that these types are often seen in combination. For instance, we might have sounds which feature a prominent harmonic spectrum, but also have some amount of distributed, noisy components, which might be a very important characteristic of that sound. For instance, a flute note will have very strong harmonics combined with some wideband noise (caused by blowing of air). Other sounds might have a great number of inharmonic components at their onset (start), which die away very quickly, leaving a stable harmonic spectrum. In this case, because the duration of the inharmonic element is short, we still perceive the overall result as pitched. However, the onset components are still very important in defining the whole object. These are actually a very common feature of many instrumental sounds, which are given the name *transients*. Figure 1.13 shows a plot of a piano note split into its harmonic, long-term, steady spectrum, and its short-lived onset transients and noise.

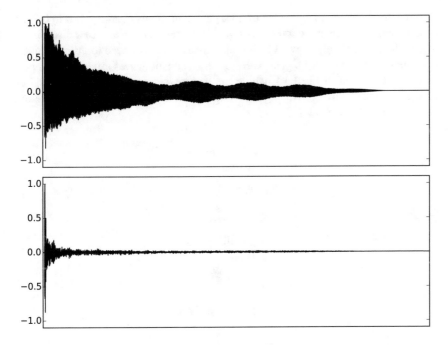

Fig. 1.13 A plot of a piano note split into steady, long-term frequency components (top) and transients/noise (bottom), from [59].

1.3 Signals in and out of the Computer

Having discussed some salient features of audio and music signals mostly in abstract terms, removed from their existence as physical phenomena, we now turn back to look at them in a more concrete way. The major question we want to examine is how these signals can exist inside a computer. This will allow us to move on to manipulate them with confidence.

The first principle we need to look into is the fundamental difference between the representation of an audio signal inside the computer and its counterpart outside it. When we speak of a wave travelling through the air (or any other medium), we have something that is *continuous*, both in time and in value. That is, if we could focus on an infinitesimal time interval, there would be no break or gap in the function that describes the signal. This is the case for waves that exist as electric signals travelling on a wire, or as a mechanical disturbance in the air, or as the vibrations in the three ossicles of our middle ear, etc. All of these forms are *analogous* to each other. Moving from one to another is a matter of *transducing* the signal.

To represent signals inside the computer, however, we need to look for an alternative form. This is mainly because the computers we normally use (*digital computers*) hold data in discrete units, instead of a varying continuous quantity[5]. So in order to capture and manipulate a continuous signal, we need to *encode* it in a discrete representation. Conversely, if we want to reproduce it, we will need to *decode* it into a continuous form. The full signal path from the acoustic waveform to the computer and back is shown in Fig.1.14, where the various stages of transduction and encoding/decoding (through Analogue-to-Digital and Digital-to-Analogue converters, ADC and DAC) are shown.

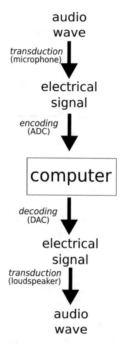

Fig. 1.14 The full signal path from acoustic wave to computer.

1.3.1 Digital signals

The types of signals we work with inside the computer are called *digital* signals. This means that the signal is discrete in time and discrete in value (or as we say more commonly, in amplitude). There are many ways that could be used to represent a continuous waveform digitally. The most common of these, which is perhaps also the conceptually simplest, is to take measurements, *samples*, of the values of the

---

[5] While *analogue* computers exist, they are not the types of machines considered in this book, even though an analogue synthesiser can be considered as a specialised form of such a computer.

waveform at regular intervals, and then assign them a discrete value within a given range, producing a series of numbers that describe the signal.

This process allows us to manipulate the waveform in many ways as if it were the original, analogue, object by conceptually thinking that each number is smoothly connected to the next one, reconstructing it as a continuous signal. However, while this is a very useful abstraction, it is also important to know that there are limits outside which this concept breaks down. The rest of this section will discuss the nature of this type of digital encoding in a more formal way, also looking at how to manage its limitations.

To derive a digital signal $s(n)$ from an analogue function $f(t)$, we start by selecting values from it at evenly spaced periods in time, T, [76]:

$$s(n) = f(nT) \tag{1.6}$$

More formally, this is the application of a *Dirac comb* III_T[11] to the function:

$$\{III_T f\}(t) = \sum_{-\infty}^{\infty} f(kT)\delta(t - kT) \tag{1.7}$$

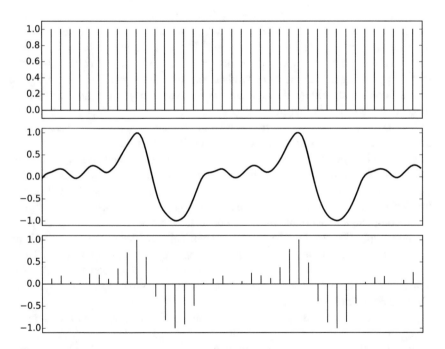

Fig. 1.15 The sampling process: we apply a Dirac comb (top) to a signal (middle), selecting the waveform values at each sampling period (bottom).

This process is known as *sampling*, and is illustrated in Fig. 1.15. The Dirac comb effectively selects values from each sampling point, which are separated by T, the *sampling period*. More generally, we will use the concept of a *sampling frequency* (or *rate*) f_s that is its reciprocal ($f_s = \frac{1}{T}$).

The Sampling Theorem [75, 94] guarantees that if our signal contains only frequencies between $-\frac{f_s}{2}$ and $\frac{f_s}{2}$, it will be recoverable in the decoding process. This means that, within these constraints, it is an accurate discrete-time representation of the continuous waveform. The range of frequencies from $-\frac{f_s}{2}$ to $\frac{f_s}{2}$ is often called the *digital baseband*, and any frequencies outside this will be represented by an *alias* inside that range. This means that components outside the baseband cannot be recovered correctly after decoding and will be mixed with the ones inside it. This is called *aliasing* or *foldover*, as frequencies are folded over the edges of the baseband, reflected back into it. The absolute limit of the digital baseband, $f_s/2$, is also called the *Nyquist* frequency.

The following expression can be used to determine an encoded frequency f_d from a continuous-signal frequency f_c, given a certain sampling frequency [68]:

$$f_d = f_c - \mathrm{int}\left(\frac{2f_c}{f_s}\right) \times f_s \tag{1.8}$$

Generally speaking, we do not need to be concerned with the encoding process, as all but the most limited ADC devices will make sure that all frequencies outside the baseband are removed before sampling. The issue, however, lies with signals generated (or modified) by computer instruments. If we are not careful, we may introduce aliasing in the process of synthesising an audio signal. For this reason, we tend to look for algorithms that are band-limited (or nearly so), which can be used to generate waveforms that fit the digital baseband. Figure 1.16 demonstrates the difference between two waveform spectra, one that was created with a band-limited process and has no aliasing, and another with substantial foldover.

Aliasing is generally undesirable, but it can of course be used creatively in some cases. We need to watch carefully because there are many situations where aliasing can creep in and possibly spoil the sound of our synthesis processes. We will be noting these conditions as they arise, when discussing the various instrument designs.

To complete the analogue-to-digital conversion, we need another step to obtain discrete amplitudes from the continuous signal. This is called quantisation [35] and involves assigning numbers, or quantisation steps, to ranges of input values. There are various ways of defining these steps and the one we will use here is called *linear* quantisation, which divides the range from minimum to maximum amplitude into equal-sized regions.

The number of regions covering the min-max range will depend on how much storage we assign to each sampled value. This is measured in *bits*, which are the basic unit in the binary number system that is used by computers. One bit can store two distinct values (e.g. 0 or 1), or two quantisation regions. If we add more bits, we can increase the number of regions (and consequently decrease their size). In the case of integers, the number of regions Q_N will be

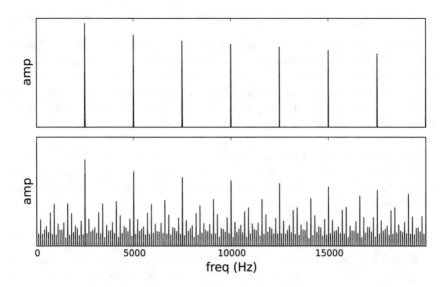

Fig. 1.16 Comparison between a band-limited (top) and aliased (bottom) waveforms.

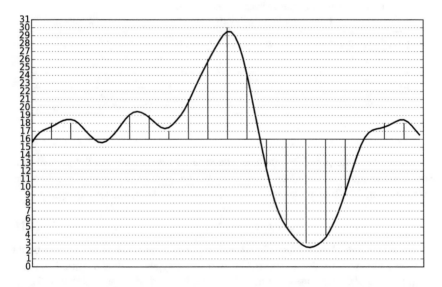

Fig. 1.17 A continuous waveform, sampled and quantised linearly into 32 regions (using 5 bits).

$$Q_N = 2^{N_b} \tag{1.9}$$

where N_b is the number of bits in each sample. In the quantisation process, all input values falling within one region are assigned to its nominal mid-range quantity. This is shown in Fig. 1.17, where we see that some of the output values are very close to the sampled values, others are not. From this we see that a certain amount of quantisation error arises from the process. This increases with the size of the region and it limits the dynamic range of the system (the range of amplitude variation that is available), as the error takes the form of noise that is added to the signal. The dynamic range is defined as the ratio between the maximum and minimum peak (or RMS) amplitudes, or as the maximum signal-to-noise ratio (SNR) [98]. This follows directly from Eq. 1.9 and we normally display this ratio on a dB-unit scale (see Section 1.2.1) for convenience:

$$SNR_{max} = 20\log_{10} Q_N \tag{1.10}$$

Table 1.2 shows the dynamic range of various sizes of integer (whole-number) samples. The SNR can be degraded if the signal does not use the full range of numbers in the system (e.g. its maximum amplitude, also called *full scale*).

Table 1.2 Dynamic range for various integer sample sizes.

N_b	dynamic range (dB)	peak (absolute) amplitude
8 (signed)	48	128
8 (unsigned)	48	256
16	96	32,768
24	144	8,388,608
32	192	2,147,483,648
64	385	9,223,372,036,854,775,807

If real (decimal-point) numbers are used to represent sampled values, we have some additional considerations [104]. Firstly, the values are encoded in a floating-point format, which is made up of a mantissa and an exponent, for example

$$s = \text{mantissa} \times b^{\text{exponent}} \tag{1.11}$$

where b is some chosen base (2, 10, e, etc.). So, in general, it is the size of the mantissa that will matter to us. In the two usual types of floating-point number, single and double precision, we have 32 and 64 bits of storage, which are divided 24:8 and 53:11 for the mantissa and exponent, respectively. In normal mode, we gain one extra bit of precision in the mantissa owing to the normalisation process [32]. So, effectively, we have dynamic ranges of 156 and 331 dB for single and double-precision floating-point numbers. An added advantage of the floating-point representation is that the total SNR is not degraded if the signal amplitude does not reach full scale. While fixed-point (integer) signals have their maximum peak amplitude determined by the precision (see table 1.2), floating-point samples are by convention set in the range of -1.0 to 1.0.

One issue to pay attention to when working with digital signals is exactly to do about keeping them within the permitted range. When a signal goes outside this range, we have *clipping*, where the edges of the waveform get cut off, causing a certain amount of amplitude distortion. In general, clipping in the input often occurs before the analogue-to-digital conversion and can be controlled by reducing the level of the signal coming into the computer. When generating or processing a waveform, it is possible to introduce digital clipping as a result of numeric overflow (attempting to use values beyond the storage capacity given by N_b). When overflow is not treated correctly, the quality of the audio can be affected dramatically. It is possible to apply soft forms of clipping that prevent the harsh audio distortion resulting from it, by gently shaping the signal as it approaches the maximum, or at least just preventing overflow. Also, when using floating-point processing, depending on the hardware, it is possible to exceed (slightly) the normal range without distortion. However, this is something that needs to be managed carefully, especially when mixing various signals, as their combination can easily push the output over the limit.

The digital encoding discussed here is called *linear pulse-code modulation* (linear PCM). It is the most common way to handle signals in the computer. There are many other forms of encoding, but they are not usually employed when we are manipulating waveforms in the way we will do in this book. They are often used as a means of re-encoding signals to improve transmission, reproduction or storage. In particular, there are various ways of *compressing* digital data that can be applied to PCM audio signals, but these encodings are applied after any synthesis or processing stages.

1.4 Conclusions

In this chapter, the most salient aspects of audio and music signals were introduced. The concept of the wave as an abstraction of a concrete physical phenomenon was put forward. The fundamental parameters of audio waveforms, namely amplitude, frequency and phase were defined first in general, and then in the context of very simple signals called sinusoids, and finally generalised in the form of the spectral representation. The special characteristics of musical signals were outlined, together with a general description of the major spectral types we will be encountering in this book.

The final section of the chapter dealt with a number of practical aspects of manipulating signals in the computer. We looked at the full path from source to destination of the audio waveforms that are sent into and out of computers. The principles of analogue-to-digital encoding and decoding were outlined, looking at the concepts of sampling and quantisation. These operations and the key parameters defining them were presented in a more formal way, which will be important in the following chapters when we start to generate and process digital audio signals.

Chapter 2
Music Programming Environments

Abstract This chapter introduces the music programming environments that will be used throughout the book. We start by giving an overview of the roles each language might play in the development of computer instruments. Following this, we focus on the three specific languages we will employ, Python, Csound, and Faust. Python can be used in its role as a glue and host language containing dedicated code written in the Csound language. It can also be employed for complete signal processing programs. Csound is a prime DSL for music, which can be used in a variety of ways. Finally, we introduce Faust as a way of programming dedicated signal-processing algorithms that can be placed within the Csound environment or used as stand-alone programs. The chapter concludes with an overview of how these three systems can coexist in an instrument development scenario

Computer music instruments have a long history of software support. It stretches all the way back to the early 1960s and the introduction by Max Mathews of his MUSIC III program for the IBM 7904 computer [63]. This was an unprecedented invention, an *acoustical compiler*, as Mathews called it, which provided the form for a multitude of computer music programming environments that were to follow. His idea, informed by earlier experiments with MUSIC I and II, was that musicians, composers, researchers, etc. would be better served by a system that had open possibilities, based on building blocks that could be connected together. In other words, rather than a program that can generate sound given parameters, one that can create new programs for that purpose [88].

His acoustical compiler was able to create software based on a given description that could be employed in music composition: a computer instrument [45]. Mathews followed it with MUSIC I [64, 100], and MUSIC V [65]. From this lineage of software, also known as the MUSIC N family, there are a number of systems that are currently in use. The one we will be employing in this book is *Csound* [59], which will be introduced in this chapter. In Csound, we have a complete domain-specific language (DSL), and a performance environment to run compiled programs that is unrivalled in terms of its breadth of sound-processing possibilities.

© Springer International Publishing AG 2017
V. Lazzarini, *Computer Music Instruments*,
https://doi.org/10.1007/978-3-319-63504-0_2

Alongside the very domain-specific programming environments offered by the MUSIC N-derived software, we should also consider the role of general-purpose programming languages (GPPLs) in the development of computer instruments. It is true to say that we could dispense completely with domain-specific approaches, because any language could be employed for this purpose. In practice, however, we will see that it is far more productive to use a DSL for specific aspects of instrument development (e.g. the signal-processing elements of it). A GPPL might not have certain components built into it that would facilitate their design and implementation.

However, there are two roles into which GPPLs could be slotted quite well. The first is to provide a glue for control and signal-processing components of our system. For example, to allow the user to create a host user interface, graphical or otherwise, to a collection of instruments. In this instance, while, again, many languages could be chosen, the most appropriate ones are the high-level scripting languages. This is because they are very straightforward to use and generally provide many possibilities for interfacing, either built in as part of the programming environment or as loadable modules. In this book, we will be employing the *Python* programming language for this purpose, as it has all the attributes listed above and, in addition to these, has excellent support for integration with Csound. Python, on its own, will serve as a means of demonstrating signal-processing algorithms, as it can provide compact code with excellent readability.

The second role in which we could find the use of a GPPL is at lower level in the implementation of computationally-intensive tasks. In this case, we are better served by a language that is classed as implementation level, such as C or C++. Specific elements of an instrument that are more demanding in terms of resources should ideally be programmed using these languages and compiled into components that can be loaded by the DSL. In this book, we will be using these languages indirectly, by employing a signal-processing language, *Faust* [77], which can be compiled into efficient C++ code, or directly into a loadable object using a just-in-time (JIT) compiler.

This chapter will first consider these three systems, Csound, Python, and Faust, separately, providing an informal overview of their languages and programming. Following this, the ways in which these can be integrated into a single workflow will be discussed. These languages will be then used in examples throughout the book.

2.1 Python

Python is a scripting language employed in a variety of applications, from user interfaces to scientific computing. It is a so-called multi-paradigm language, allowing a diverse variety of programming approaches. It is generally simple to use and allows for quick prototyping of ideas. Since it relies on an *interpreter* to translate programs, it tends to be slow for computationally-intensive applications. However, it is very easily extendable, which allows it to use pre-built components to speed up some tasks.

The system (language/interpreter) has witnessed a major version change in recent years. It has moved from version 2 (which was kept at 2.7) to 3 (currently 3.5) in a non-backward compatible way. This has confused many users, as it took a long time for many popular third-party components to be upgraded. At the time of writing, Python 3 has become well established with most of the important dependencies being made available for it. In this book, we will be adopting this version. This section will introduce the key language elements that will allow the reader to follow the program examples in later chapters. Further information can be found in the official language reference documentation at python.org.

Python can be used interactively by typing code directly into an interpreter shell, or by supplying it with programs written in text files. For instance the following command starts an interactive session:

```
$ python3
Python 3.5.1 (v3.5.1:37a07cee5969, Dec  5 2015,
21:12:44)
[GCC 4.2.1 (Apple Inc. build 5666) (dot 3)] on darwin
Type "help", "copyright", "credits" or "license" for
more information.
>>>
```

At this stage, commands can be typed at the prompt (>>>). Programs can also be passed as files to the command:

```
$ python3 helloworld.py
Hello World!!
```

In this case, the file `helloworld.py` consists of a single line:

```
print('Hello World!!')
```

Python is a dynamically-typed language. This means that variable types are defined according to the data they are carrying and the type can be re-defined for the same name, if we need to:

```
>>> a = 'hello'
>>> type(a)
<class 'str'>
>>> a = 1
>>> type(a)
<class 'int'>
```

In this example the variable a was first holding a text string, and then a whole number, which represent two distinct types. In addition to the various types we will discuss below, Python has a special one, None, which indicates the absence of a value and is used in situations when we need to specify this explicitly.

Built-in numeric data in Python is divided into three basic categories: integers (<class 'int'>), floating-point numbers[1] (<class 'float'>) and complex numbers (<class 'complex'>). Integers have unlimited precision and floating-point data is implemented using double precision. Complex numbers have two parts, real and imaginary, both of which are floating-point. A complex constant has the form a + bj, where the letter j is used to label the imaginary part.

In addition to these, a special type of numeric data are the boolean numbers. These are used in truth expressions and can assume two values: True or False. These are a subtype of integers and behave like 1 and 0, respectively. When converted to text, they become the strings 'True' and 'False'.

Arithmetic statements are very similar to various other languages. Assignment is defined by =, and the common arithmetic operators are as usual: -, +, *, /, and \% minus/difference, addition, multiplication, division, and modulus, respectively. To these, Python adds ** for exponentiation and @ for matrix multiplication. In Python 3, division always returns a floating-point number even if we are using integer operands. To force truncation, we can use a double slash (//). As usual, expressions can be combined together with parentheses (( )) to determine the precedence of operations. Some of these operations can also have different meanings in contexts that are not arithmetic, as we will see at various times in this book.

One significant aspect of the Python syntax is its insistence on indentation levels as code block delimiters. This typographical approach works well for code readability, but it is awkward in interactive use. For instance, to define a callable subroutine, we use the the keyword def followed by the routine name, a list of parameters, and a colon. The code that is going to be executed by this subroutine has to be placed at a higher level of indentation to indicate that it belongs to the subroutine:

```
def  mix(a,b):
    c  = a+b
    return c
```

All the code that belongs together in a block needs to be at the same level of indentation. This is also the case in other syntactic constructions that we will see later in this chapter. Finally, comments starting with a hash (#) and continuing to the end of the line are allowed.

2.1.1 Control of flow and loops

As with other languages, control of flow can be achieved with a combination of if – elif – else statements, which evaluate a truth expression. Code in sections defined by these keywords needs to be formatted at a higher indentation level. For example:

```
if a < 0:
```

---

[1] The floating-point data type is used to represent real numbers in programs.

```
    print('negative')
elif a == 0:
    print('zero')
else:
    print('positive')
```

As usual, the logical expressions following the `if` and `elif` keywords can have multiple terms, combined by use of the boolean logical operators `and` and `or`.

The basic looping construction in Python is defined by `while`, followed by a truth expression and a colon. The loop block is then placed, indented, under this:

```
n = 0
while n < 10:
    print(n)
    n += 1
```

Here, we can optionally place an `else` section after the loop body, which will execute when the expression is not true.

Another form of loop can be found in the `for ... in ...` statement. This is slightly more complex in that it involves the concept of an *iterable* object that is used to construct the loop. This type of loop is designed to work with sequences such as lists and strings. Here is an example of this construction:

```
for x in [0,1,2,3,4]:
    print(x)
```

which, when executed, yields

```
0
1
2
3
4
```

We can get gist of it from this example. The variable `x` takes the value of a different member of the list each time the body of the loop is executed.

2.1.2 Sequences

In Python there are a number of data types that can be sequentially organised and manipulated, which are known generically as *sequences*. Among these we have lists, text strings, tuples and ranges.

Lists

A list is a sequential collection of data, which can be of the same or different types. Lists are mutable, i.e. they can be enlarged, shortened, have items deleted, replaced, inserted, etc. To create a list we can use square brackets as delimiters, or use the named constructor:

```
l = [0,1,2,3,4]   # list with 5 items
k = []            # empty list
k = list()        # using the named list constructor
```

Lists can be accessed using a similar square-bracket notation. With it, we can take slices of the list, as well as individual items:

```
>>> a = [0,1,2,3,4]
>>> a[0]
0
>>> a[2:4]
[2, 3]
>>> a[0:5:2]
[0, 2, 4]
>>> a[:5:2]
[0, 2, 4]
>>> a[1:]
[1, 2, 3, 4]
>>> a[2] = 100
>>> a
[0, 1, 100, 3, 4]
```

The general form for the slice is [start:end:increment]. In addition to this, there are a number of operations that can be used to insert, append, copy, delete, pop, remove, reverse, and sort a list. In addition to these, we can also have *list comprehensions* of the following form:

```
>>> [x*x for x in [1,2,3,4]]
[1, 4, 9, 16]
```

where an expression involving a variable (such as x*x) can be iterated over a list input to produce an output list. This is a concise way to create a list based on the evaluation of a function or expression.

Tuples

Tuples are immutable[2] sequence types that can be used to store data objects together. Access is similar to that for lists, and slices are also possible. However, none of the mutation operations are available for tuples. They can be created using a comma-separated sequence of items, or using the named constructor:

```
l = 0,1,2,3,4              # tuple with 5 items
k = tuple([0,1,2,3,4])     # using the named constructor
```

---

[2] This means that they cannot be changed.

Ranges

The range type is also immutable and represents a sequence of integers with a start and an end. Ranges can be constructed with the forms `range(end)` or `range(start,end,increment)`, where the end value is not actually included in the sequence. To make a list out of a range it is necessary to create one using a list *constructor* (`list()`):

```
>>> list(range(5))
[0, 1, 2, 3, 4]
>>> list(range(0,5,2))
[0, 2, 4]
>>> list(range(1,6))
[1, 2, 3, 4, 5]
```

Text strings

One special case of a sequence is a string. This is another immutable type, which is used to hold text characters. Strings can be single quoted, embedding other double quoted strings; double quoted, embedding other single-quoted strings; or triple quoted, containing multiple lines:

```
>>> a = 'this is a "string"'
>>> a
'this is a "string"'
>>> a = "'this' is a string"
>>> a
"'this' is a string"
>>> a = '''these are lines
... lines and
... lines'''
>>> a
'these are lines\nlines and\nlines'
>>> print(a)
these are lines
lines and
lines
```

A number of operations can be performed on strings. For instance, the addition operator (+) can be used to concatenate strings:

```
>>> a = 'hello'
>>> b = ' world!!'
>>> a+b
'hello world!!'
```

Strings can be formatted in ways similar to the standard C library function `printf` using the % operator. For instance:

```
>>> a = "this is a float: %f and this is an integer: %d"
>>> a % (1.1, 2)
'this is a float: 1.100000 and this is an integer: 2'
```

2.1.3 Functions

Functions, an example of which has already been presented earlier in this chapter, can be used to implement subroutines or proper functions, with or without side-effects. The basic syntax is:

```
def  function-name(argument-list):
     function-body
```

The function body has to be indented at a higher level to make it a separate block. Alternatively, if the function consists of one line only, then it can be place on the same line after the colon. The comma-separated argument list can have initialisers, or default values, for parameters, and the function body can include the keyword `return` to return results:

```
def  add_one(a=0):
     return a+1
```

Functions can return multiple values as tuples:

```
def  add_one_two(a=0):
     return a+1,a+2
```

On a function call, arguments are passed by *object reference*. This means that a name on the argument list refers to a data object[3] that is passed at the function call and it can then be used in the function body accordingly. For instance:

```
def  func(a):
     print(a)
```

The name a refers to whatever object was given to the function and then is passed to `print`. This might be of different types and the function will adjust its behaviour accordingly:

```
>>> func('hello')
hello
>>> func(1)
1
```

---

[3] All data in Python are objects of one type or another: numbers, sequences, text strings, etc. are examples of built-in types. Other types can be user-defined, or supplied by packages.

Arguments can also be passed out of order with keywords, which default to the argument names:

```
>>> def func(a,b): print(a+b)
...
>>> func(b='world!!', a='hello ')
hello world!!
>>>
```

The scope of the names defined in a function is local, or within the function body or block. On assignment, an object is created and bound to a name, locally. If we want to keep it, we have to return a reference to it:

```
def   func(a):
        b = a + 1 # an object is created and assigned to b
        return b  # the object reference is returned
```

However, it is possible to access objects in the global namespace directly from within a function, but only if the given name is not used in the function:

```
>>> def func(): return b + 1
...
>>> b = 2
>>> func()
3
>>> b
2
```

We can also use a special keyword `global` to refer explicitly to a global name. In any case, we should avoid directly accessing objects that are in the global scope.

2.1.4 Classes

Python supports the creation of new types of objects with dedicated operations. This is done through class definitions:

```
class NewType:
    ...
```

Classes are, by default, derived from a basic type called `object`, unless we explicitly pass another type in the declaration:

```
class NewType(ParentType):
    ...
```

Following the class declaration, we normally place a number of methods to operate on this new type. These are special types of functions that are bound to the class and can manipulate it:

```
class NewType:
    def method(self):
        ...
```

The first argument to a class method is always the class object (by convention, we call it `self`) and can be used to refer to all the class members (variables and functions, or *methods*). In addition to this argument, we can have a normal argument list. Accessing methods of a given instance of a class use the syntax `object.method()`, where `object` is the variable name of a given class, and `method` one of its member functions. Classes can also have data members, or instance variables, and these can be created by methods, added to the class definition or even added to a new object after its creation. These can also be accessed using a similar syntax, `object.variable`.

An object of a newly created class can be created by invoking its constructor, which is a function with the name of the class:

```
a = NewType()
```

Constructors can be declared as a method named `__init()__`. This is used to perform custom initialisation tasks:

```
class NewType:
    def __init__(self,a):
        self.a = a
    def print(self):
        print(self.a)
```

In this case, the constructor allows us to create objects with different attributes:

```
>>> a = NewType(1)
>>> a.print()
1
>>> a = NewType(2)
>>> a.print()
2
```

Inheritance allows a derived class to access methods and variables in the parent class as if they were its own. A specific method in the base class can be accessed using the syntax `Parent.method()`, if necessary.

2.1.5 Modules and libraries

Python can be extended with a variety of specialised modules or libraries. Normally these are available from the central repositories at python.org or through package management systems[4]. Once a module/package is installed, we can load it using the `import` statement:

```
import package_name
```

From there, all code in a package `pkg` can be accessed using the form `pkg.obj`, where `obj` stands for any method, variable, constant, etc. defined in the package. We can also give an alias `name` to a package:

```
import package_name as name
```

Pylab

In this book, we will be using a number of external libraries. In particular, we will employ the scientific libraries SciPy, NumPy, and matplotlib, packaged together as pylab. This will be commonly invoked by the command:

```
import pylab as pl
```

Most of the key SciPy/NumPy/matplotlib objects will be exposed through this interface. Occasionally, we might need to load specific parts of these libraries separately, if they are not available through pylab. The fundamental data type that is used by all of these libraries is the NumPy `array`, which has been designed to optimise numeric computation. It behaves very similarly to other sequence types (such as lists), but it actually holds a numeric array in native form (integers, float and complex subtypes are provided).

2.2 Csound

Csound is a sound and music computing system that can be used in a variety of applications. It is programmed and run via its DSL through a variety of frontend software. In this section we will introduce the basic aspects of its operation and the main features of its language. We will concentrate on the most relevant elements that will allow the reader to follow the examples in the next chapters. Further concepts will be introduced as required. A complete reference for the system and language can be found in [59].

The Csound language is compiled into an efficient set of signal processing-operations that are then performed by its audio engine. All operations to compile and run code can be accessed through a high-level programming language interface, the Csound Application Programming Interface (API). We will first look at the most salient aspects of the language before discussing how to use the API (through Python) to run the system.

---

[4] Many operating systems have their own package managers for adding new components. In addition, Python has pip (https://pip.pypa.io/en/stable/), which is used specifically for Python packages on multiple platforms.

2.2.1 The language: instruments

The basic structuring unit in Csound is the *instrument*. This is used to hold a series
of operators called *unit generators* that are defined by various *opcodes* in the system. An instrument is a model of a sound-processing/generating object that can be
instantiated any number of times. An instrument is defined as follows:

```
instr ID
...
endin
```

where `ID` is an integer or a name identifying the instrument. Inside instruments
we place unit generators that can process audio. For example, the opcode `oscili`
generates a sine wave signal with a specified amplitude and frequency (the first
and second parameters). This opcode takes frequencies in cps (Hz) and amplitudes
within the range specified by Csound. This is set by the constant `0dbfs`, which determines the maximum full-scale peak amplitude in the system (see section 1.2.1).
It defaults to 32,768 (for backwards compatibility), but it can be set to any positive
value. To get audio out of this instrument, we use the opcode `out`:

```
instr 1
  out(oscili(0dbfs/2,440))
endin
```

To run this instrument, we need to schedule it, passing the instrument name, the
start time and the duration to `schedule`

```
schedule(1,0,1)
```

The call to `schedule` is placed outside instruments in global space. This call is
executed once Csound starts to perform. It could also be placed inside an instrument,
in which case it would recursively schedule itself.

2.2.2 Variables and types

Csound has a number of variable types. These define not only the data content, but
also the action time. When an instrument is run, there are two distinct action times:
initialisation and performance. The former is a once-off run through the code that is
used to initialise variables and unit generators, and execute code that is designed to
run once per instance. Performance time happens when instruments process signals,
looping over the unit generator code.

In the example above, at init time, the oscillator is allocated and its initialisation
routine is executed. At perf time, it produces the audio signal that is sent to the
output. Signals can be of two types: scalars and vectors. The first type is used for
control operations (e.g. parameter automation) and the second for audio. The two

Table 2.1 Basic variable types in Csound.

variable type	action time	data format
i	init	scalar
k	perf	scalar
a	perf	vector

action times and the two types of signals yield three basic variables types: i, k and a (see Table 2.1).

Types are defined by the first letter of the variable names: i1, kNew, asig, etc. We can rewrite our first example using these:

```
instr 1
   iamp = 16000
   icps = 440
   asig = oscili(iamp, icps)
   out(asig)
endin
```

In this case, the first two lines will execute at i-time, followed by the initialisation of the oscillator. At performance time the two last lines will be run in a loop for the requested duration, filling and consuming audio signal vectors. The size of these vectors is determined by the system constant ksmps and defaults to 10. The sampling rate is also set by the constant sr and its default value is 44,100. A a third quantity, the control rate (kr) is set to $\frac{sr}{ksmps}$, determining how often each control (scalar) variable is updated. This also determines the processing loop period (how often the processing code is repeated).

The scope of variables is local to the instrument. Global variables can be created by attaching a 'g' to the start of the variable name (e.g. gi1, gkNew, or gasig). Instruments can contain external initialisation arguments. These are identified by parameter numbers $p1...pN$. The first three are set by default to the instrument number, start time and duration (in seconds, with -1 indicating indefinite duration). The following arguments can be freely used by instruments. For instance:

```
instr 1
   iamp = p4
   icps = p5
   asig = oscili(iamp, icps)
   out(asig)
endin
```

This will allow us to use parameters 4 and 5 to initialise the amplitude and frequency. In that case, we can have sounds of different pitches:

```
schedule(1,0,1,0dbfs/3,440)
schedule(1,0,1,0dbfs/3,550)
schedule(1,0,1,0dbfs/3,660)
```

Control variables are used to make parameters change over time. For instance, we could use a trapezoidal envelope generator (linen) to shape the amplitude of our example instrument, cutting out the clicks at the beginning and end of sounds:

```
instr 1
   idur = p3
   iamp = p4
   icps = p5
   ksig = linen(iamp,0.1,idur,0.2)
   asig = oscili(ksig, icps)
   out(asig)
endin
```

With these three basic types, it is possible to start doing signal processing with Csound. More advanced types will be introduced later when we need them to design certain instruments.

There are two alternative syntaxes in Csound. The examples above demonstrate one of them, where opcode parameters are passed as comma-delimited lists inside parentheses and the output is put into variables using the assignment operator (=). The other form uses an output–opcode–input format, with no parentheses and no explicit assignment, only spaces:

```
instr 1
   idur = p3
   iamp = p4
   icps = p5
   ksig linen    iamp,0.1,idur,0.2
   asig oscili   ksig, icps
        out      asig
endin
```

The two ways of writing instrument code can be combined together. In particular, the function-like syntax with the use of parentheses can be used to inline opcodes in a single code line:

```
asig oscili linen(iamp,0.1,idur,0.2),icps
```

When opcodes are used with more than one output, however, this syntax is not available, and we need to employ the more traditional output-opcode-input format.

Csound code can be run directly from its various frontends. The most basic one is the csound command, to which code can be supplied as a text file. For instance, if our instrument and schedule lines above are written to a file called *instruments*, we could run it in this way:

```
$ csound --orc instruments -o dac
```

where the -o dac option tells Csound to play the sound directly to the soundcard. Alternatively, we could use the API to compile and run programs directly.

2.2.3 *The API*

The Csound API provides complete access to the system, from configuration to code compilation and audio engine performance. It is based on a C/C++ set of functions and classes, but it has been wrapped for various languages. In the case of Python, there are two alternative packages that can be used, csnd6 and ctcsound. The former is only available for Python 2 and the latter can be used in both versions 2 and 3.

ctcsound

The ctcsound package provides full access to the Csound C API, in addition to a few extra components. It can be loaded with the command:

```
import ctcsound
```

A typical use of the package involves the following steps:

- Creating a Csound object
- Setting up options to control the compiler and engine
- Compiling code
- Starting and running the audio engine
- Interacting with the compiled instruments

For example, the following code creates a Csound object, sets it to output audio to the soundcard, compiles some code and performs it:

Listing 2.1 A simple Python Csound API example.

```
import ctcsound

code = '''
instr 1
    idur = p3
    iamp = p4
    icps = p5
    ksig = linen(iamp,0.1,idur,0.2)
    asig = oscili(ksig, icps)
    out(asig)
endin

schedule(1,0,1,0dbfs/3,440)
schedule(1,1,1,0dbfs/3, 550)
schedule(1,2,1,0dbfs/3, 660)
'''

cs = ctcsound.Csound()
cs.setOption('-odac')
```

```
cs.compileOrc(code)
cs.start()
cs.perform()
```

When running the Csound engine, it is often simpler to start a new thread for performance, so that the user can interact with it. For this purpose, we have the CsoundPerformanceThread class, which takes a Csound object and allows it to be played:

Listing 2.2 Using a performance thread

```
cs = ctcsound.Csound()
cs.setOption('-odac')
cs.compileOrc(code)
cs.start()
perf = ctcsound.
  CsoundPerformanceThread(self.cs.csound())
perf.play()
```

Methods for pausing, stopping and joining the thread are available (pause(), stop(), and join()). With Csound running in a separate thread, we can start new instruments running and set controls in them using methods of the Csound class. Both of the above examples (with and without threads) can be used to run the Csound code examples in this book, with Python as the host environment. In this case all we need to do is to replace the triple-quoted string in code by the program in question. We might also need to set other options, which we can do by calling setOption() more than once with the option as an argument. A full working example demonstrating these possibilities will be shown in section 2.4.

The software bus

Communication with a running Csound object is achieved via the software bus. It works through logical *channels* that can be used to transmit control or audio data to and from the host. The principle is straightforward. A channel with any given name can be declared in Csound with the opcode chn_k or chn_a for control or audio, respectively:

```
chn_a(Sname, imode)
chn_k(Sname, imode)
```

where Sname is a string with the channel name and imode is the mode (1, 2, 3 for input, output, and input/output respectively). If a channel is not declared, it is created when it is used for the first time.

From Csound, we can write and read with the following opcodes

```
xs chnget Sname
chnset xs, Sname
```

where xs is either an i-, k-, or a-type variable containing the data to be written (chnset) or to where we should read the contents of the channel (chnget).

The Python host uses the following methods of a Csound object to access a channel:

```
# get a value from a channel
Csound.controlChannel(name)
# set a value of a channel
Csound.setControlChannel(name,val)
# get a signal from an audio channel
Csound.audioChannel(name, samples)
# put a signal in an audio channel
Csound.setAudioChannel(name, samples)
```

The software bus can be used throughout a performance and it is particularly useful for the design of user interfaces for real-time instruments (see section 2.4 for a complete example).

Scheduling events

A host can schedule events in a running object in the same way as we have used a schedule() opcode inside Csound code. This can be done via the inputMessage() method of either the Csound or the CsoundPerformanceThread classes:

```
Csound.inputMessage(event)
CsoundPerformanceThread.inputMessage(event)
```

The argument event is a string containing a message that starts with the letter 'i' (for 'instrument'). Parameters are separated from one another by empty spaces (not commas, since they are not function arguments, but items of a *list*). For instance, a Csound code line schedule(1,0,-1,0.5,400) can be translated into the following event string:

```
event = 'i 1 0 -1 0.5 400'
```

In this case, we are telling Csound to instantiate instrument 1 immediately (p2 = 0) for an unlimited duration (p3 = −1, *always on*) passing 0.5 and 400 as parameters 4 and 5, respectively. With both scheduling of events and software bus channels, we cancreate complete graphical user interface instruments with Csound and Python, as demonstrated in Sect. 2.4 and in Appendix B.1.

2.3 Faust

Faust [77] is a system designed to manipulate audio streams, and with which we can implement a variety of digital signal processing algorithms. Programs created in Faust can be more efficient than equivalent code written in other high-level music programming languages. These are generally used as components for specialised purposes in host environments (such as Csound), although stand-alone applications can also be developed. Programs can also be used to generate graphic flowcharts that can assist with the design and implementation of algorithms.

Faust code describes a process on an audio stream in a purely functional form. The following minimal example demonstrates thd basic principle:

```
process = + ;
```

Faust statements are terminated by a semicolon (;). The arithmetic operator (+) takes two inputs, mixes them and produces one output. It is implicit in this operation that it takes two inputs and produces one output. Similarly, an example of a minimal program with one input and one output is given by

```
process = * (2);
```

where multiplication by 2 is used. This is a function that maps one input to one output.

2.3.1 Stream operators

Faust has a number of specialised operators to work on streams. Sequential (:) and parallel (,) operators can be used to organise the order of function application. For instance:

```
process = _ ,2 : *;
```

takes audio input (using _) and the constant 2 at the same time and applies them in a sequence (:) to the function '*'. The (_) is a placeholder for a signal.

Two other important operators are split (<:) and merge (:>), for sending one signal into parallel streams and for mixing two inputs into one stream, respectively. The following example,

```
process =   _ <: *;
```

squares a signal, and

```
process = _,_:> _;
```

mixes two streams.

The '@N' delay operator applies a delay of N samples to a signal, with which we can create an echo effect:

```
process = _<:_@4410,_:> _;
```

Similarly, the single quote and double quote operators apply one- and two-sample delays, respectively. The following program,

```
process = _<: _ +_' : * (0.5);
```

applies averaging to consecutive samples in an input signal.

Complementing these operations, Faust includes the concept of *feedback*, denoted by the '~' (tilde) operator. This is used to feed the output of a stream back into its input.:

```
process = _ + _ ~ _;
```

To the left of the tilde we have the input(s) to the operator. The first one is always the feedback signal, and so if we want to add an input, we need to provide a second one, which in this case is mixed with the feedback. To the right of the tilde we have a feedback path that implies a 1-sample delay. On this side we can place other operations if needed (e.g. scaling, futher delays, etc.). A simple example is given by

```
process =    @(4409)+_ ~ *(0.9);
```

which will mix an input signal with its feedback delayed by 4409 samples. The feedback signal is attenuated by gain (0.9). The total delay is 4410 samples (4409 + the 1-sample delay from the feedback), which at $f_s = 44100$ (the default sampling rate) equals 0.1 sec. This process implements an echo with feedback (multiple repetitions of the input signal). All of these Faust operations are illustrated with flowcharts demonstrating each process in Fig. 2.1. These were generated directly from the Faust process definitions.

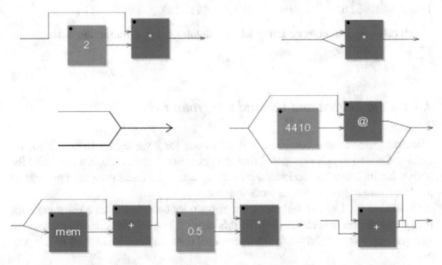

Fig. 2.1 Faust operators. Top row: parallel-sequential (scaling by 2); split (squaring). Mid row: merge (mix); delay (@). Bottom row: one-sample delay (single quote); and feedback (tilde).

2.3.2 Functions

Functions can be defined to conveniently describe an operation. For instance, we could rewrite the averaging process in a clearer form as

```
f(x) = (x + x')/2;
```

which can be used at various places in a program. Anonymous (lambda) functions can also be used using the following notation:

```
\(x).((x + x')/2);
```

This can be used to simplify the notation of some algorithms, when the direct use of stream operations can make the code look cluttered.

2.3.3 Controls

Faust also allows the building of generic user interfaces (UIs). These can take different forms depending on the actual platforms on which the code can be compiled. As an example, a generic horizontal slider is defined by

```
freq = hslider("frequency", 440, 100, 1000, 1);
```

which takes as parameters a name label, a default value, a minimum, a maximum, and a minimum step.

2.3.4 Compilation and Csound integration

Faust programs can be compiled into C++ code for a variety of *architectures*, including Csound plugin opcodes. From the generated source code, we can build the library binaries that can be loaded into the system. The Faust program can then be used in Csound instruments as a new opcode.

Alternatively, Csound includes a set of opcodes designed to take Faust code directly to compile and run it [48]. In this case, it is possible to integrate the two systems intp a very tightly-knit arrangement. There are three basic steps in this process:

1. Compilation: Faust code is compiled into a DSP factory object. This is a binary representation of the processing code that can be instantiated.
2. Instantiation: from a DSP factory object, we can create a process instance, which is ready to be run and to perform the desired signal processing.
3. Performance: with an instance in memory, it is possible to invoke its `compute()` to produce the audio output.

The Faust Csound unit generators, or opcodes, have been designed to perform these three operations in separate steps at initialisation and performance time. Faust code compilation runs at initialisation time. It does not involve any signal processing, and it does need to be repeated. More than one instance of the same Faust program can then be activated at performance time.

The performance code is subdivided into two parts, signal processing and controls, and is split into two separate opcodes. The signal processing element picks up input signals (if defined) processes them and outputs thm. Parameter control is performed by other opcodes, which can manipulate given UI components defined in the program.

Faustcompile

The `faustcompile` opcode calls the just-in-time (JIT) compiler to produce a DSP process from a Faust program. It will take a Faust program from a string. The operation can be controlled by various arguments. In Csound, multi-line strings are accepted, using {{ }} to enclose the string and S-type variables can be used to hold them.

```
ihandle faustcompile Scode, Sargs
```

```
Scode   – a string containing a Faust program.
Sargs   – a string containing Faust compiler arguments.
```

```
ihandle faustcompile "process=+;", "-vec -lv 1"
```

Faustaudio

The `faustaudio` opcode creates and runs a compiled Faust program. It works with code compiled with `faustcompile`:

```
ihandle,a1[,a2,...] faustaudio ifac[,ain1,...]
```

```
ifac – a handle to a compiled Faust program, produced by faustcompile.
ihandle – a handle to the Faust DSP instance, which can be used to access its controls with faustctl.
ain1,... – input signals
a1,... – output signals
```

```
ifac faustcompile "process=+;", "-vec -lv 1"
idsp,a1 faustaudio ifac,ain1,ain2
```

Faustctl

The `faustctl` opcode is used to access UI controls in a Faust DSP instance. It will set a given control in a running faust program:

faustctl idsp,Scontrol,kval

Scontrol – a string containing the control name
dsp – a handle to an existing Faust DSP instance
kval – value to which the control will be set.

```
idsp,a1 faustgen {{
gain = hslider("vol",1,0,1,0.01);
process = (_ * gain);
}}, ain1
faustctl idsp, "vol", 0.5
```

Faust programming is particularly useful in the prototyping and implementation of new DSP algorithms for computer instruments. Its close integration with Csound makes the three-language cooperation introduced in this chapter particularly powerful.

2.4 Programming Environment and Language Integration

At this point, it would be useful to look at how the interaction between these languages can be employed in a real-world situation. Taking Python as a host language, Csound as the audio engine, and Faust to implement a specialised audio processing algorithm, we will study the integration of these three systems introduced in this chapter. The different levels of embedding of these systems are shown in Fig. 2.2. The final section of this chapter will examine one application case study as an example of this type of integration. Note that the discussion will be steered towards the overall application design. The specific aspects of signal processing and instrument development will be explored in more detail in Part II.

2.4.1 The application

Our case study will focus on a variation of the Shapes program that is provided as one of the Python Csound API examples in the source code tree (under examples/python/shapes.py). This will be adapted to use Python 3 and the ctcsound package. Fig. 2.3 shows the main application window.

From a high-level, the program works as a synthesiser with a two-dimensional control. Once the user clicks on the circle, it changes colour from black to read and

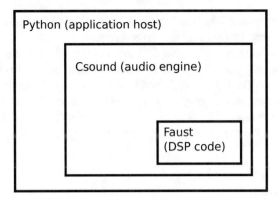

Fig. 2.2 Levels of embedding of Python, Csound, and Faust.

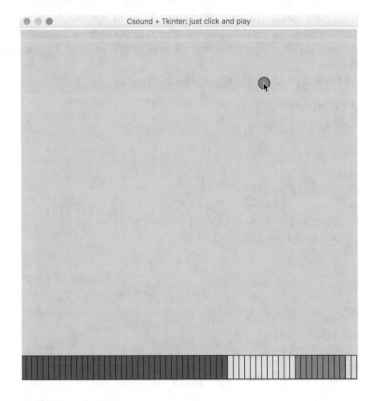

Fig. 2.3 The Shapes application.

we start hearing a sound. The graphical user interface (GUI) is designed as a canvas space to be explored by dragging the red ball around.

We can soon identify that the vertical axis controls the volume (or amplitude) of the tone, and that pitch lies in the horizontal direction. We also notice that, on clicking, there is a short *attack* followed by a long decay. When the ball is let go, the sound also continues through a shorter *release* period, in which it gets duller and softer until it stops. The bottom of the window shows a VU meter that gets illuminated when the tone is played, changing with increases in volume.

2.4.2 Instruments

The Shapes application is composed of two elements as far as its code is concerned. One of them structures the program itself, and the other controls the audio engine. The former is written in Python, while the latter is in the Csound language, containing the instrument that makes the sound we hear, which includes a process implemented in Faust.

The full Csound code for this application is contained in a Python text string:

Listing 2.3 Csound instrument code.

```
code = '''
0dbfs=1

gar init 0
gifbam2 faustcompile {{
        beta = hslider("beta", 0, 0, 2, 0.001);
        fbam2(b) = *~(((\(x).(x + x'))*b)+1);
        process = fbam2(beta);
}}, "-vec -lv 1"

instr 1
 ival = p4
 kp chnget "pitch"
 kp tonek kp, 10, 1
 kv chnget "volume"
 kv tonek kv, 10
 ain oscili 1, p5*kp, -1, 0.25
 ib, asig faustaudio gifbam2,ain
 kb1 expsegr ival,0.01,ival,27,0.001,0.2,0.001
 faustctl ib,"beta",kb1
 asig balance asig, ain
 kenv expsegr 1,20,0.001,0.2,0.001
 aenv2 linsegr 0,0.015,0,0.001,p4,0.2,p4
 aout = asig*aenv2*kenv*0.4*kv
```

```
      out    aout
  gar += aout
endin

instr 100
  a1,a2 reverbsc gar,gar,0.7,2000
  amix = (a1+a2)/2
    out amix
  ks rms gar+amix
  chnset ks, "meter"
  gar = 0
endin

schedule(100,0,-1)
'''
```

From the sounds we hear we might deduct that instrument 1 is responsible for the tone, also because the other one does not output any audio. So we can start by concentrating on that part of the code. The design of the instrument is centred around the FBAM2 algorithm[5] [50], implemented as a Faust program:

```
gifbam2 faustcompile {{
        beta = hslider("beta", 0, 0, 2, 0.001);
        fbam2(b) = *~(((\(x).(x + x'))*b)+1);
        process = fbam2(beta);
}}, "-vec -lv 1"
```

Csound produces a sinusoidal signal (a cosine wave) with its oscili opcode, which is fed into the FBAM2 process. The rest of the instrument is made up of envelopes, one of which will be used to control a timbral parameter in the Faust code, and two others to make a linear attack and an exponential decay. Every time the instrument is triggered, these envelopes will shape the amplitude and timbre of the output sound.

The sources for the amplitude and pitch parameters are the "volume" and "pitch" channels (the chnget opcode). As we have seen in Sect. 2.2.3, software bus channels such as these are used to carry control data from the Python host to Csound. On the Python side, these are accessed by the setControlChannel() method of the Csound class, as we will see in later in the code examples.

An extra effect has been added to enhance the sound in this code. In instrument 100, we have a reverb (reverbsc) to modify the sound generated by instrument 1. This effect instrument is scheduled to be running continuously, always on. Another purpose for it is to produce the RMS value that is in the VU display. This is sent to the Python host via another software bus channel:

```
instr 100
  a1,a2 reverbsc gar,gar,0.7,2000
```

---

[5] We will study this algorithm in detail later on, in Chap. 5.

```
amix = (a1+a2)/2
   out amix
ks rms gar+amix
chnset ks, "meter"
gar = 0
endin
```

Notice that we are also using a global variable (gar) to carry the audio signal from the source to the reverb. We need to initialise it at the top:

```
gar init 0
```

and then use it in instrument 1:

```
gar += a2
```

These ideas should give a sense of how instruments can be developed from the perspective of signal processing operations. There is scope for significant sound design work, which will be explored in the next chapters of this book.

2.4.3 Application code

In addition to DSP, we should also consider how we can interact with and drive the instruments we create. This is where Python and its hosting capabilities come into play. The Shapes program uses the Tkinter package, which is a standard GUI framework for Python, based on the Tk library:

```
import tkinter
```

Tkinter has capabilities to set up an application that will create one or more windows, place widgets (the graphical components such as buttons, sliders, text boxes, etc.), and manage user interaction with the application.

To use it, a program will have to:

1. Create a new class definition that derives from the tkinter.Frame base class.
2. Add any code needed to create and draw the widgets in this class.
3. Instantiate the class.

In addition, since we are embedding a Csound engine in this application, we will need to create and manage an instance of the audio engine using the API classes and methods. We can take advantage of the architecture that Tkinter imposes on the source code and place all Csound functionality in the newly created class.

Before we look into designing the application class, let's look at what we need. These are the GUI components we see in Fig. 2.3:

- A single main window frame
- A blank canvas
- A circle/ball that can be dragged around and changes colour

- A rectangle with subdivisions that change colour in response to the sound volume

Looking at the functionality that Tkinter offers, we see that the `Canvas` object might be the basic widget we are looking for. It implements an open space where structured graphics can be drawn and manipulated. So, to implement what we have described above, the application class will need to:

- Create a basic frame.
- Instantiate a Canvas.
- Draw a circle, and bind it to mouse controls.
- Draw a series of small rectangles that together look like a big one with subdivisions.

Once these things are done, we can create a Csound object and start it running. Each click on the circle will trigger a new event in instrument 1 (see Listing 2.3). We will then hear a sound, which will continue if we hold the mouse button down, and stop if we let it go.

Dragging the ball around will change the pitch and amplitude, so we will have to respond to that as well (by setting the relevant bus channels). Finally, when the application is closed, we qill have to stop the engine. This is all the application needs to implement.

Most of the work is done inside the new class we are creating. Let's call it `Application`, derive it from `tkinter.Frame`, as required, and then initialise it by following our list of tasks outlined above:

Listing 2.4 Application init code.

```
class Application(tkinter.Frame):
 def __init__(self,master=None):
  self.master = master
  self.master.title('Csound + tkinter: '
    'just click and play')
  # create main frame window
  tkinter.Frame.__init__(self,master)
  self.pack()
  # create GUI components
  self.createCanvas()
  self.createCircle()
  self.createMeter()
  # create Csound engine
  if self.createEngine() is True:
   # meter drawing
   self.drawMeter()
   # set up close button callback
   self.master.protocol('WM_DELETE_WINDOW',
                        self.quit)
   # start the application main loop
```

```
    self.master.mainloop()
    # if the engine could not start, quit
    else: self.master.quit()
```

After initialising the top-level frame, we create the GUI components, the Engine and start the application. We can look at each of these steps one at a time now. They are all methods of the Application class. The canvas creation method creates a square space, with a violet background, inside the main frame (its parent). For each Tk widget created, we need to call its pack() method for the geometry manager to place it correctly under its parent.

Listing 2.5 Canvas creation.

```
def createCanvas(self):
    self.size = 600
    self.canvas = tkinter.Canvas(self,
                                 height=self.size,
                                 width=self.size,
                                 bg="violet")
    self.canvas.pack()
```

To create a circle, we call a shape-drawing method from Canvas and place it on a variable (circle). We then bind to this circle three mouse actions: button pressing, button 1 (held) motion, and button release. Each action will trigger a callback method associated with it (which we will examine later):

Listing 2.6 Circle creation.

```
def createCircle(self):
  circle = self.canvas.create_oval(self.size/2-10,
                                   self.size/2-10,
                                   self.size/2+10,
                                   self.size/2+10,
                                   fill="black")
    self.canvas.tag_bind(circle, "<ButtonPress>",
                         self.play)
    self.canvas.tag_bind(circle, "<B1-Motion>",
                         self.move)
    self.canvas.tag_bind(circle, "<ButtonRelease>",
                         self.stop)
```

The meter is just a series of rectangles (10 of them), which are created on the canvas, and are initially coloured grey. Once Csound starts playing, the rectangles will be drawn with different colours:

Listing 2.7 Meter creation.

```
def createMeter(self):
    iw = 10
    self.vu = []
    for i in range(0, self.size, iw):
```

```
self.vu.append(self.canvas.create_rectangle(i,
        self.size-40,i+iw,self.size,fill="grey"))
```

The final set-up method creates the engine. We instantiate a Csound object, compile the code, and set up two input channels for controls. Then we create a performance thread for this engine, and start it running with its Play() method:

Listing 2.8 Engine creation.

```
def createEngine(self):
 self.cs = ctcsound.Csound()
 self.cs.setOption('-odac')
 res = self.cs.compileOrc(code)
 if res == 0:
  self.cs.setControlChannel('pitch', 1.0)
  self.cs.setControlChannel('volume', 1.0)
  self.perf =
   ctcsound.CsoundPerformanceThread(self.cs.csound())
  self.perf.play()
  return True
 else:
  return False
```

The drawing of the meter is done at regular intervals. In order to enable this, we use the after() function to schedule a call to the drawing method at the end of its definition, so that it will call itself repeatedly. This method checks for the value of a meter channel sent by Csound and fills each rectangle accordingly, with different colours:

Listing 2.9 Meter drawing.

```
def drawMeter(self):
 level = self.cs.controlChannel("meter")
 level *= 16000
 cnt = 0
 red = (self.size/10)*0.8
 yellow = (self.size/10)*0.6
 for i in self.vu:
  if level > cnt*100:
   if cnt > red:
    self.canvas.itemconfigure(i, fill="red")
   elif cnt > yellow:
    self.canvas.itemconfigure(i, fill="yellow")
   else:
    self.canvas.itemconfigure(i, fill="blue")
  else:
    self.canvas.itemconfigure(i, fill="grey")
  cnt  = cnt + 1
 self.master.after(50,self.drawMeter)
```

Now we can examine the mouse action callbacks. When the mouse button is pressed, we will issue the `Play()` method, which will locate the current widget being handled by this user `event` and issue an `inputMessage()` to the engine (via its performance thread). This method takes a string with a list of parameters, starting with an identifier code ('i' for instrument event), and followed by each parameter separated by spaces:

```
'i 1 0 -1 0.5 440'
```

which should be read: instr 1, starting now (0), for an unlimited time (-1), with p4 = 0.5 and p5 = 440. As we have seen before, a negative p3 can be used to make a sound event of indefinite duration:

Listing 2.10 Starting a sound.

```
def play(self,event):
    note = event.widget.find_withtag('current')[0]
    self.canvas.itemconfigure(note, fill='red')
    self.perf.inputMessage('i1 0 -1 0.5 440')
```

Stopping a sound uses a very similar method, except that when we issue an event to Csound, we need to send a negative p1:

Listing 2.11 Stopping a sound.

```
def stop(self,event):
    note = event.widget.find_withtag('current')[0]
    self.canvas.itemconfigure(note, fill='red')
    self.perf.inputMessage('i-1 0 1 0.5 440')
```

When moving the mouse to drag a circle, we will need to do two things: (a) control Csound and (b) move the circle by tracking the mouse position. The canvas is the widget to which the user event is registered. We can get the canvas coordinates of the event, and as before get the item (circle) we are manipulating. By calling the `coords` method of the `Canvas` class, we can move the item anywhere inside the canvas. Finally, we just set the control channels in the Csound engine according to the position of the circle:

Listing 2.12 Moving the circle.

```
def move(self,event):
    canvas = event.widget
    x = canvas.canvasx(event.x)
    y = canvas.canvasy(event.y)
    item = canvas.find_withtag("current")[0]
    canvas.coords(item, x+10, y+10, x-10, y-10)
    self.cs.setControlChannel("pitch", 2.0*x/self.size)
    self.cs.setControlChannel("volume",
                      2.0*(self.size-y)/self.size)
```

To complete the class, we need to implement a `quit()` method that will stop the performance thread, so that Csound can exit cleanly. To do this, we stop the thread, which will also close the engine, and join it until it finishes, and then we destroy the GUI objects:

Listing 2.13 Cleaning up.

```
def move(self,event):
def quit(self):
  self.perf.stop()
  self.perf.join()
  self.master.destroy()
```

The main program then becomes just a matter of instantiating the application class:

```
Application(tkinter.Tk())
```

The full Python code for this program can be found in Appendix B (Listing B.1). Further details on the development of user interfaces will be explored in Chap. 8.

2.5 Conclusions

In this chapter, we introduced the three programming languages that will be used throughout this book. We discussed the most salient details, exploring aspects of each language that will be relevant in the next chapters. It was demonstrated how these three systems can be integrated into a smooth working environment for instrument development.

It is true that Python, Csound and Faust could potentially be used on their own for this purpose. In fact, in this book we will be employing them separately at certain times. In particular, we will also use Python to describe and demonstrate aspects of signal processing that are relevant to our discussion.

However, it is good to be able to point out the advantages of a multi-language approach to programming. Each one of the systems introduced here has a particular strength: Python with its simple syntax and wide range of facilities for application hosting; Csound, as a leading DSL for music and audio; and Faust with its support for efficient implementation of new DSP algorithms. The combination of these qualities provides for a very rich computer music experience.

Part II
Synthesis and Processing

Chapter 3
Source-Filter Models

Abstract This chapter will discuss the key aspects of instrument development that fall into the category of source-filter models. Commonly placed under the general label of subtractive synthesis, these methods include different ways to define sound generation sources, with different spectral characteristics, and the many types of filters that can shape these. The text starts by looking at various types of broadband generators for periodic and noise signals, followed by an exploration of dynamic parameter shaping. The next section discusses filters in terms of their fundamental characteristics and their application in spectral modification. The chapter concludes with instrument examples demonstrating source-filter synthesis.

Source-filter models are often known by the misnomer *subtractive synthesis*, which tries to capture the principle behind these techniques, namely the shaping of a complex signal, but uses the wrong terminology. The fact is that subtraction is used only in an abstract sense here, and it is better to speak of modification of source signals via specialised processors called *filters* (as well as through envelope and modulation control of synthesis parameters). The overall idea is that we start with a spectrally-rich input and then dynamically shape it to match a desired sound.

In this chapter we will look at the principles involved in modelling sound through this source-filter arrangement. We will start by complementing, in a more formal way, some of the ideas about spectra that were introduced in Chap. 1. Following this, we will look at different types of source signal generation, and ways in which we can shape synthesis parameters. This is completed by a study of the different filter types and their applications. Programming examples are shown throughout the text to provide context for the reader.

3.1 Sound Spectra

When we introduced the concept of the sound spectrum in section 1.1.4 as composed of a combination of pure frequencies, each one a sinusoid with a different amplitude,

© Springer International Publishing AG 2017
V. Lazzarini, *Computer Music Instruments*,
https://doi.org/10.1007/978-3-319-63504-0_3

frequency and phase, we did not define it in a formal way. There are various ways of doing this and one of them is to present it as a generic mix of signals. This can be useful up to a certain point, but it becomes unwieldy if we have too many of these frequencies to mix.

However, let's try to express this idea more formally so we can understand it better. A complex arbitrary sound signal $s(t)$ can be defined as a sum of partials given by [40]

$$s(t) = \sum_{k=1}^{N} A_k \cos(\omega_k t + \theta_k) \qquad (3.1)$$

where A_k are the amplitudes, $\omega_k = 2\pi f_k$, the frequencies (f_k defined in cps, or Hz)[1], and θ_k, the phases of each component. This is an additive synthesis approach. To demonstrate how this works, we can use a small Python script to generate a signal and plot its waveform:

Listing 3.1 Additive spectrum generation.

```
sr = 44100.
end = sr/50
pi = pl.pi
t =  pl.arange(0,end)
partials = [(1.,100.,-pi/2),
            (0.5,300.,0),
            (0.25,500.,-pi/2),
            (0.1,700.,0),
            (0.05,800.,0),
            (0.01,900.,-pi/2)]
s = pl.zeros(end)

pl.subplot(211)
pl.title("separate partials")
pl.xlim(0,end)
for n in partials:
        a,f,ph = n
        p = a*pl.cos(2*pi*f*t/sr + ph)
        s += p
        pl.plot(t,  p)

pl.subplot(212)
pl.title("waveform")
pl.xlim(0,end)
pl.plot(t,  s)
pl.tight_layout()
```

---

[1] Throughout this book, t is the time of sample n, $\frac{n}{f_s}$, where f_s is the sampling rate. This will allow us to keep the formulation simple without having to refer to f_s constantly.

```
pl.show()
```

This simple program generates a signal consisting of a sum of six frequencies, with amplitudes, frequencies and phases defined in the `partials` list of tuples. The time axis (in samples) is given by a NumPy array constructed using `pl.arange()` between 0 and the end, and the pylab function `zeros` creates an empty array to receive the mix of partials[2]. These are plotted both separately and as a waveform mix (Fig. 3.1). It is possible to see that, in this case, all frequencies (100, 300, 500, 700, 800, 900) fall in the harmonic series of $f_0 = 100$ Hz, and thus we will most likely perceive this sound with a pitch at that fundamental. Note also that some partials are cosines ($\theta_k = 0$) and some are sines ($\theta_k = -\frac{\pi}{2}$).

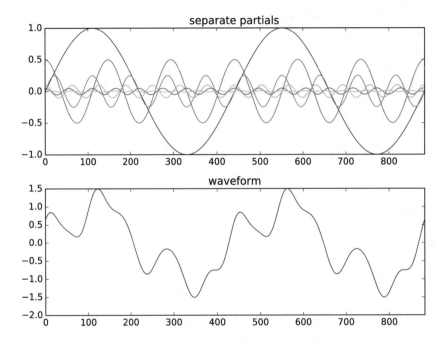

Fig. 3.1 Plot of a waveform (bottom) and its components (top), as generated by the program in Listing 3.1.

It is not necessarily the case for all spectra that frequencies are aligned with a given harmonic series. Many sounds have inharmonic spectra, where a harmonic series is not found. More generally, we should also expect the amplitudes A_k and frequencies f_k to be functions of time in their own right, changing the spectrum dynamically. This would characterise the additive synthesis approach more completely.

---

[2] This will be a common practice in the Python synthesis examples throughout this book.

For the moment, it is useful to think of sound sources that are periodic as in this example, with a static set of harmonic partials. This simplifies the synthesis process significantly and since there are many musical sounds that can be modelled this way, it is a good place to start. In this case we have

$$s(t) = \sum_{k=1}^{N} A_k \cos(\omega_0 k t + \theta_k) \tag{3.2}$$

where $\omega_0 = 2\pi f_0$ is our fundamental frequency, and all components are multiples of it. This is also known as the Fourier series, to which we will come back in Chaps. 4 and 7. We could simplify it further by fixing the phase shift θ to be the same for all partials. The following program does this, generating a square wave spectrum:

Listing 3.2 Square wave generation.

```
import pylab as pl

sr = 44100.
end = sr/50
pi = pl.pi
t =  pl.arange(0,end)
f = 100
amps = [1.,0,1./3,0,1./5,0,1./7,0,1./9.]
ph = -pi/2
s = pl.zeros(end)

pl.subplot(311)
pl.title("harmonics")
pl.xlim(0,end)
k = 1
for a in amps:
    p = a*pl.cos(2*pi*f*k*t/sr + ph)
    s += p
    k += 1
    pl.plot(t, p)

pl.subplot(312)
pl.title("square wave")
pl.xlim(0,end)
pl.plot(t, s)

pl.subplot(313)
pl.title("spectrum")
end = len(amps)+1
pl.xlim(0,end)
pl.ylim(0,1.1)
pl.stem(pl.arange(1,end),amps,markerfmt=" ")
```

```
pl.tight_layout()
pl.show()
```

We can see that the square wave is made up of a sum of sine waves ($\theta = -\frac{\pi}{2}$), whose even harmonics are missing and whose amplitudes are set to $\frac{1}{k}$, where k is the harmonic number. The program plots the separate components, the resulting square wave and its magnitude spectrum (the amplitudes of each harmonic), as shown in Fig. 3.2.

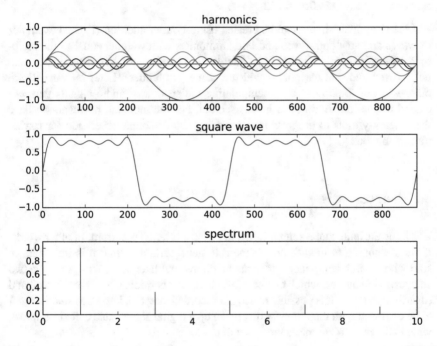

Fig. 3.2 The plot of a square waveform (middle) and its components (top) and magnitude spectrum (bottom), as generated by the program in Listing 3.2.

Turning to Csound, we can do this very easily with a function table generator (GEN). There are over forty of these built into the language and three of them allow us to create waveforms from harmonic parameters. The simplest is GEN 10, which takes only the harmonic amplitudes (or weights), as in the amps list used in the program above. GENs write to a (function) table, which is an array of floating-point numbers that can be accessed by various types of opcodes, such as oscillators. In the example below, we create the same square wave as in Listing 3.2 into function table 1 using the ftgen opcode[3] which takes a table number, time of generation,

---

[3] A complete and up-to-date reference for this and all other opcodes discussed in this book can be found at http://csound.github.io/docs/manual.

size, GEN number and parameters. The instrument uses the table in an oscillator to play a 440 Hz tone:

```
gi1 = ftgen(1,0,16384,10,1,0,1/3,0,1/5,0,1/7,0,1/9)

instr 1
  out(oscili(p4,p5,gi1))
endin

schedule(1,0,2,0dbfs/2,440)
```

Likewise, other shapes can be created using Fourier addition. Typical examples of waves are: sawtooth (even and odd harmonics, with weights equal to the reciprocal of the harmonic number ($\frac{1}{k}$); triangle (odd harmonics, $\frac{1}{k^2}$, cosine phase); and pulse (even and odd harmonics with the same amplitudes). Later on we will see other ways of looking at and manipulating the spectra of audio signals that will extend the notions introduced in this section. For the moment, however, we have the necessary tools to progress into exploring source-sound generation for further transformations.

3.2 Sources

The basic attribute that a source signal must have is a rich spectrum. In other words, if we are hoping to modify and shape it to get a certain output, it is important to start with enough components that can be filtered out later. In general, we have two categories of sources, which can be mixed together: periodic signals and broadband (distributed) noise. It is possible to use pre-recorded material, known as *samples* (not to be confused with the numbers that make up a digital signal, also called samples), which allows us to tap into a larger set of source sounds.

3.2.1 Periodic signal generators

The basic source generator for periodic sounds is the oscillator, which we have already used in some examples without defining it properly. The oscillator is designed to output a signal that repeats over time. In some cases, it can also be used to generate a one-shot function, whose period extends over the whole duration of a sound segment. Typically, an oscillator is made up of three components: (a) function lookup; (b) phase increment; and (c) scaling. The following Python example demonstrates the idea, generating a 1-second signal:

Listing 3.3 Sine-wave oscillator.

```
import pylab as pl
```

```
def sinosc(amp, freq, ph, sr):
    incr = freq*2*pl.pi/sr
    # 1 sec of audio
    out = pl.zeros(sr)
    for i in range(0, sr):
      # function lookup
      out[i] = pl.sin(ph)
      # increment
      ph += incr
    # scaling
    out *= amp
    return out
```

Of course, in this case we are only generating a sine wave and we need something with more components. The function used for this can be anything that will generate a single cycle of a waveform. In fact, we will typically pre-calculate this in a memory block called a *table* and then read from it. This is known as *function-table lookup*:

Listing 3.4 Table-lookup oscillator.

```
def oscil(amp, freq, ph, tab, sr):
    incr = freq*len(tab)/sr
    out = pl.zeros(sr)
    for i in range(0, sr):
        # table lookup & modulus
        out[i] = tab[int(ph)%len(tab)]
      # increment
      ph += incr
    # scaling
    out *= amp
    return out
```

In this case, given that the function is periodic over the table length, the increment has to be scaled appropriately as

$$i = f_0 \times \frac{N}{f_s} \tag{3.3}$$

where N is the table length and f_s, the sampling frequency. We also need to make sure the table reading is kept within bounds (for which in this example we use the Python modulus operator %, which is defined for both positive and negative integers). It is possible to calculate the increment recursively. The idea is to keep the phase θ normalised in the range $0 \leq \theta < 1$ by taking its value modulo 1 (or its fractional part):

Listing 3.5 Table-lookup oscillator with recursive phase calculation.

```
def phasor(ph, freq, sr):
    incr = freq/sr
    ph += incr
```

```
    return ph - floor(ph)

def oscil(amp,freq,ph,tab,sr):
    out = pl.zeros(sr)
    for i in range(0,sr):
      ph = phasor(ph,freq,sr)
      out[i] = tab[int(ph*len(tab))]
    out *= amp
    return out
```

In fact, the phase increment operation is such a fundamental component that it is often offered separately from the oscillator. In the case of Csound, this is provided in the phasor opcode. The phase is effectively a signal that keeps ramping from 0 to 1 at a rate given by the frequency parameter. In Faust, we find that the recursive algorithm shown above in Listing 3.5 can be implemented in a particularly compact expression:

Listing 3.6 Phase increment.

```
mod1(a) = a - floor(a);
incr(freq) =  freq / float(SR);
phasor(freq,ph) =  incr(freq) : (+ : mod1) ~ _ :
                 +(ph) : mod1;
```

Here the increment is summed recursively with its decimal part. An initial phase ph is added to the signal and the modulo 1 function is applied again after this offset to keep it within bounds. A flowchart for this algorithm is shown in Fig. 3.3.

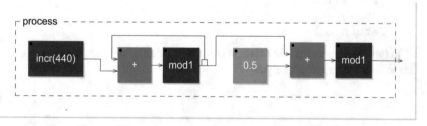

Fig. 3.3 Flowchart of recursive phase calculation algorithm, with freq = 440 and ph = 0.5.

Another issue that is important to consider is the precision of the table lookup. In the example in Listing 3.4, this is not ideal because we are *truncating* the table index to look up a sample from the table. This will lead to a distortion in the shape of the function, compromising the quality of the signal. We can improve this by using an interpolated lookup, which attempts to find a closer value for the function when the index is not integral. The simplest case is linear interpolation, which models the in-between values as straight lines between the calculated table positions:

Listing 3.7 Interpolating oscillator.

```
def oscili(amp,freq,ph,tab,sr):
    size = len(tab)-1
    out = pl.zeros(sr)
    for i in range(0,sr):
        ph = phasor(ph,freq,sr)
        pos = size*ph
        poi = int(pos)
        frac = pos - poi
        out[i] = tab[poi] + frac*(tab[poi+1] - tab[poi])
    out *= amp
    return out
```

In this case, we need an extra point at the end of the table called a *guard point*, to interpolate the final position. This is reflected in the code by the fact that we are assuming that the actual size of the function table is one position longer than the number of points in the calculated function. The guard point can be either a copy of the first position in the table, for cases when the oscillator will be wrapping around the ends of the table, or an extension of the contour of the function, for when we will be using the oscillator to read only once through the table (in the one-shot applications mentioned before). Higher orders of interpolation can also be used (linear interpolation is order 1). Third order (cubic) interpolation is another common choice. In Csound, the opcodes `oscil`, `oscili`, and `oscil3` are commonly used for truncated, linear, and cubic lookup, respectively.

Function tables should ideally be band-limited. This is because we would like to avoid the aliasing noise that can easily be introduced by the oscillator if we are not careful. Once a signal is generated with aliasing, it is very hard to suppress it. For this reason, we should avoid simple functions based on simple geometric shapes, for example based on piecewise linear segments. Instead we should try if at all possible to approximate our desired waveform by using a Fourier series, as discussed in section 3.1.

Using tables with pre-defined numbers of harmonics will allow us to avoid aliasing, as we know what the highest component in a signal is. The simple calculation is to make sure that $f_0 \times N \leq \frac{f_s}{2}$, where N is the highest harmonic number. This poses one particular problem, which is that if our range of fundamental frequencies is wide, low pitches might sound too thin as they will lack higher components (which were limited by our aliasing considerations). A solution is to use multiple tables, which can be organised into octave, or sub-octave ranges, and the oscillator is designed to switch between them according to its frequency.

In Csound, the `vco2` opcode uses this strategy. It creates a set of function tables from a choice of different spectrum types (sawtooth, square, triangle, pulse, and user-defined), and selects these according to the fundamental frequency. The output signal is guaranteed to be alias free, with the full bandwidth from the fundamental to the Nyquist frequency. The following program in Listing 3.8 runs a minimal instrument with `vco2` producing a sawtooth wave (using its default mode), and plots the result (Fig. 3.4).

Listing 3.8 The opcode `vco2` producing a band-limited sawtooth wave.

```
import ctcsound as csound
import pylab as pl

code = '''
instr 1
 out(vco2(p4,p5))
endin
schedule(1,0,1,0dbfs,100)
'''

cs = csound.Csound()
# the option -n turns off audio output
# as we are just plotting it
cs.setOption('-n')
cs.compileOrc(code)
cs.start()
spout = cs.spout()
sig = pl.zeros(cs.ksmps()*88)
n = 0
for i in range(0,88):
    # this runs one perf cycle at a time
    cs.performKsmps()
    for i in spout:
        sig[n] = i/cs.get0dBFS()
        n+=1

x = pl.arange(0, len(sig))/cs.sr()
pl.figure(figsize=(8,3))
pl.xlabel("time (s)")
pl.tight_layout()
pl.plot(x,sig)
pl.show()
```

The different modes allow us to generate other band-limited wave shapes, as shown in Fig. 3.5, using the same function table method. In fact, `vco2` allows users to draw arbitrary geometric shapes, which can be supplied through the auxiliary opcode `vco2init` as a set of function tables for band-limited signal generation. The original shape is analysed and its Fourier series determined. From this, tables are constructed so as to be band-limited according to the various ranges of fundamental frequencies.

This is demonstrated by the following Csound code:

```
iN = 16384
```

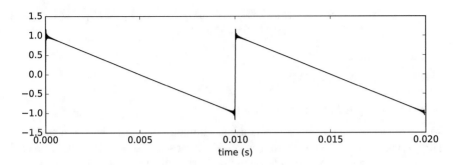

Fig. 3.4 A band-limited sawtooth wave, as generated by the program in listing 3.8.

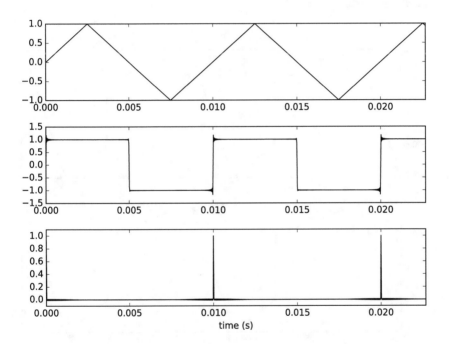

Fig. 3.5 Three band-limited waveforms from vco2. From the top: triangle, square and pulse.

```
isf = ftgen(1,0,iN,7,-1,iN/4,1,iN/2,0.5,1,-0.5,iN/4-1,-1)
ifn = vco2init(-1, 100,1.05,-1,-1,isf)

instr 1
  out(vco2(p4,p5,p6))
endin
schedule(1,0,1,0dbfs,100,14)
```

where function table 1 contains a non-band-limited geometric shape (created by
GEN 7, which draws functions from piecewise linear segments), and the `vco2init`
opcode creates a set of band-limited tables derived from it for `vco2` (using mode 14,
user-defined shapes) to generate a clean signal. We see a comparison between the
band-limited signal from `vco2` with the output from an ordinary oscillator reading
directly from table 1 (Fig. 3.6).

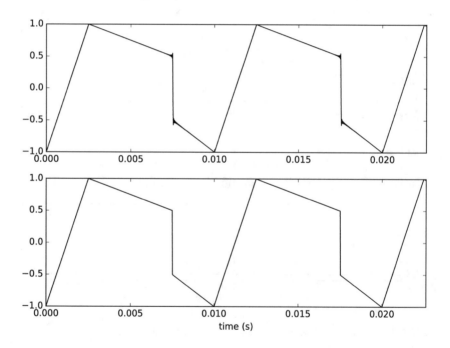

Fig. 3.6 Comparison between a band-limited waveform from `vco2` (top) and its non-band-limited
original form (bottom).

The mechanism of table selection according to the fundamental frequency in
`vco2` is also exposed for use with ordinary opcodes (such as `oscili`, for instance).
Working alongside `vco2init`, we have `vco2ift` and `vco2ft`, which can be
used for this purpose:

```
iN = 16384
```

```
isf = ftgen(1,0,iN,7,-1,iN/4,1,iN/2,0.5,1,-0.5,iN/4-1,-1)
ifn = vco2init(-isf, 100,1.05,-1,-1,isf)
gibas = -isf

instr 1
 itab = vco2ift(p5,gibas)
 out(oscili(p4,p5,itab))
endin
```

Similarly, we can look for other wave shapes to derive a set of band-limited function tables from. Although it is slightly awkward, it is possible to read audio from a file and use it a source for this process. There are some limitations, however: (a) the table sizes in the set need to be the same; (b) the original table is restricted to have a power-of-two length; and (c) vco2init imposes a limit on the table size (2^{18}). The beginning and end of the source sound need to be crossfaded for a continuous tone, which also requires the use of a specialised looping oscillator, flooper or flooper2.

The following example demonstrates these ideas. We load part of a file called "flutec3.wav", using GEN 1, which reads from soundfiles and places them in a function table of size 2^{16}, which at $f_s = 44,100$ means about 1.48 seconds of audio. Then we create a set of band-limited tables of the same size (iN) based on this source sound. To play it back at the correct pitch we need to know what the original f_0 is, if at all possible. In this case, we are told it is a C at 523.25 Hz. The playback pitch is the ratio of the fundamental and this base frequency. This is actually equivalent to the phase increment in Eq. 3.3, which determines how fast the oscillator proceeds through the table. So, for table selection we need to supply the correct frequency for this phase increment, multiplying it by the ratio $\frac{f_s}{N}$ (where N is the table length):

```
Sname = "flutec3.wav" // source sound
gicps = 523.25         // original pitch
iN = 2^16              // table size
gisf = ftgen(1,0,iN,1,Sname,0,0,0,1)
ifn  = vco2init(-gisf, 100,1.05,iN,iN,gisf)
gibas = -gisf

instr 1
 ip = p5/gicps
 ifr = ip*sr/ftlen(gisf)
 itab = vco2ift(ifr,gibas)
 ixfd = 0.25
 ilpd = ftlen(itab)/sr
 a1 = flooper2:a(p4,ip,0,ilpd,ixfd,itab)
   out(a1)
endin
schedule(1,0,5,0dbfs,440)
```

The `flooper2` opcode takes a loop duration (`ilpd`), set to the table size, and a crossfade time[4]. These parameters can be adjusted to get an optimal continuous-tone loop. While a little more elaborated, this allows the pitch of the sound to be transposed upwards with no aliasing. In the case of a more limited transposition, and with sounds that do not have very strong high-frequency components, it may be tolerable to play directly from the original function table, as very little or no aliasing will be heard in the output. This would allow us to simplify our instrument and accept tables of any size, which will be able to accommodate the whole duration of source sounds. For this, we just give the table a length of 0 and GEN 1 takes care of creating a table that is long enough to hold the whole soundfile:

```
gisf2 = ftgen(ifn,0,0,1,Sname,0,0,0,1)
instr 2
 ip = p5/gicps
 ixfd = 0.25
 ilpd = ftlen(gisf2)/sr
 a1 = flooper2:a(p4,ip,0,ilpd,ixfd,gisf2)
    out(a1)
endin
schedule(2,0,5,0dbfs,600)
```

These signal sources are often called *sampled-sound* generators, as they rely on pre-recorded sounds. Given that these can be of any spectral type, they are not strictly limited to periodic waveforms, although our examples have looked at this particular type. More generally, we can incorporate a variety of different sounds as sources for further modification, including various types of noise, which will be explored in the next section. Finally, it should also be noted that there are other ways of generating band-limited or quasi-band-limited (alias-suppressed) signals for further shaping. We will have a look at these in Chap. 4, as they involve more specialised methods of signal processing.

3.2.2 Broadband noise and distributed spectra

Complementarily to the periodic sources discussed in the previous section, broadband noise plays a very important role in modelling spectra that cannot be created by oscillators or similar unit generators. As discussed before, in Chap. 1, the main characteristic of noise is that it is not made up of distinct partials, but of components spread over a certain band in the spectrum. At the extreme, we have the full-band *white* noise, whose energy distribution is flat across all frequencies.

---

[4] You might have noticed the syntax `flooper2:a()`. This is needed because the opcode can also output more than one channel at a time (e.g. stereo). Since we are using the mono version, we need to indicate the return type (a single audio variable, a).

We can observe this by generating a random sequence as our signal and plotting its waveform and spectrum as we did in the previous sections. The Csound code for this is very simple:

```
instr 1
 out(rand(p4))
endin
schedule(1,0,2,0dbfs)
```

and the plots are shown in Fig. 3.7.

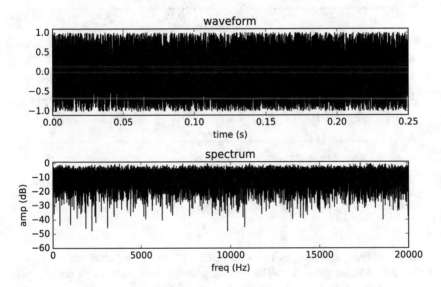

Fig. 3.7 White noise waveform and its amplitude spectrum, as generated by rand.

As we can see from this, it is not very useful to look directly at the spectrum magnitudes, as they do not indicate the presence of discretely-defined partials. In such a case we should look for the overall shape of the spectrum, which we can call the *spectral envelope*. This is a far more useful way of describing different types of noise, as we then have a higher-level view of their energy distribution. In the case of the spectrum in Fig. 3.7, the spectral envelope is flat across all frequencies (any slight fluctuation you might observe in it is just a by product of the finite-time nature of the analysis we are doing).

There are several different types of noise generators whose output exhibits a generally flat spectrum, but whose waveforms have different audible characteristics from white noise. These are mostly to do with the grain characteristics of the particular random distribution used, and the difference is be more evident in the waveform plot. Compare, for instance, the plots of two waveforms generated using Gaussian and Cauchy noise distributions in Fig. 3.8, whose spectral envelopes are

also flat, with the white noise in Fig. 3.7. These can be generated with the opcodes
`gauss` and `cauchy` in Csound.

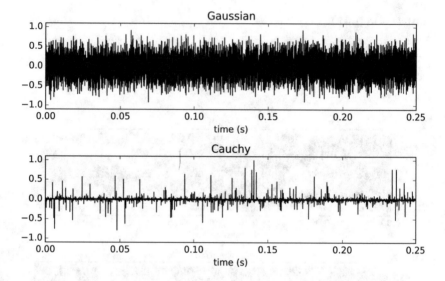

Fig. 3.8 Gaussian and Cauchy noise waveforms, as generated by the `gauss` and `cauchy` opcodes
in Csound.

It is possible, however, to generate different types of spectral envelopes with
band-limited noise generators. The general principle is to start with a random generator and then select samples of its output at a lower rate than the sampling frequency. We can do this in more than one way. For instance, we can hold the random
value until we pick a new one. This is a type of sample-and-hold operation, and it is
performed by the opcode `randh`:

```
instr 1
 out(ranh(p4,p5))
endin
schedule(1,0,2,0dbfs,2500)
```

In this example, we draw a new random number 2500 times a second. We can see
it in Fig. 3.9 that its spectral envelope is not flat anymore and that it has more energy
in the lower part of the spectrum. The bumps are determined by the frequency of
the sample-and-hold operation, with troughs at its multiples and peaks in between
them.

If instead, we connect the random points with straight lines, interpolating between them, we will get even more suppression of higher frequencies. This is what
`randi` does, using similar parameters to `randh`. A plot of the resulting waveform
and spectral envelope is shown in Fig. 3.10

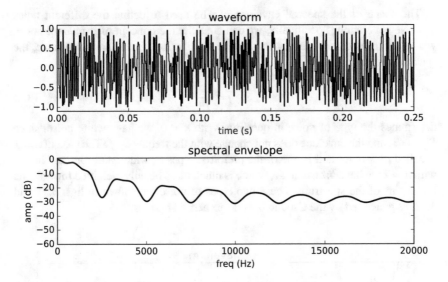

Fig. 3.9 Waveform and spectral envelope of band-limited noise, as generated by `randh`.

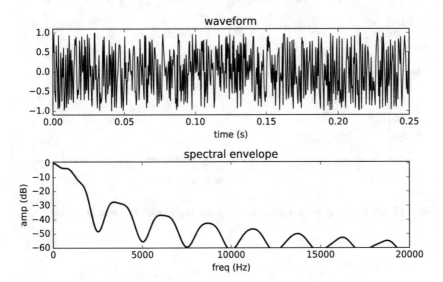

Fig. 3.10 Waveform and spectral envelope of band-limited noise, as generated by `randi`.

The shape of the spectral envelope can be used to define the different types of noise generation. Specifically, we can consider the concept of *fractional* noise, whose nature is determined by its spectral distribution. From this perspective, the spectral envelope

$$S(f) = \frac{1}{f^n} \qquad (3.4)$$

determines the type of noise in question. With $n = 0$, we have white noise, since $S(f) = 1$, and the envelope does not change with the frequency f. The case of $n = 1$ is called *pink* noise and is one of the preferred types of wide-band source. Finally, with $n = 2$ we have *brown* noise, which is much more heavily weighted towards the lower part of the spectrum. These two types of noise are shown in figs. 3.11 and 3.12, as generated by the Csound opcode `fractalnoise`.

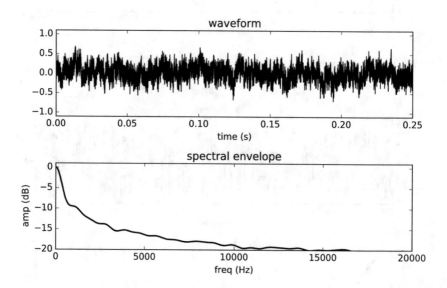

Fig. 3.11 Waveform and spectral envelope of $\frac{1}{f}$ noise, as generated by `fractalnoise`.

3.3 Dynamic Parameter Shaping

So far, we have been looking at sound generation in a static way, as we have examined the types of spectra we can produce. However, one of the most important aspects of source-filter models is the possibility of dynamically changing the characteristics of the signal. This can be as simple as modifying the volume of the sound

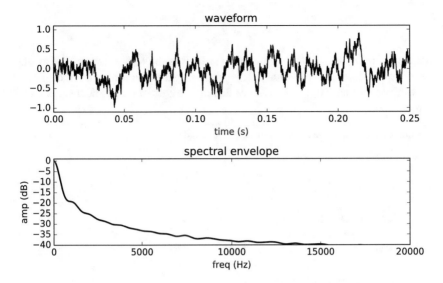

Fig. 3.12 Waveform and spectral envelope of $\frac{1}{f^2}$ noise, as generated by `fractalnoise`.

source, but it can also involve the shaping of any parameter available in an instrument. There are two basic ways to do this, which can be used separately or in combination, through the use of envelopes and modulation. The distinction between these can be subtle, but in general the former involves the application of a non-periodic, single-shot function of time, whereas the latter has to do with periodic fluctuations of parameters.

3.3.1 Envelopes

Envelopes are functions of time that are used to control a parameter. In Chap. 1, we introduced this concept with an example of n amplitude envelope. It is true that the most common use of envelopes is in the shaping of the peak amplitude of a signal, to model the fact that sounds do not normally start and end with the same volume. However, we can speak of a more general concept that can be applied to other aspects of an audio signal.

The amplitude envelope of a signal can be estimated to give us an idea of how this parameter evolves for different types of sound. We can do this by calculating the root-mean-square (RMS) amplitude of the signal and track how it changes over time. This can be defined as

$$\sqrt{\frac{1}{N}\sum_{n=0}^{N-1} s(n)^2} \tag{3.5}$$

for a signal $s(n)$ of length N. In other words, it is the square root of the average of the squared samples of a signal. It is proportional to the peak amplitude, but not necessarily the same value. We can see this for the case of a sinusoidal signal in this program:

Listing 3.9 Computing the RMS amplitude of a sinusoid.

```
import pylab as pl

def rms(sig):
    s = 0.0
    for samp in sig:
        s += samp**2
    return pl.sqrt(s/len(sig))

t = pl.arange(0,1000)
print(rms(pl.sin(2*pl.pi*t/1000.)))
```

from which we get 0.707 as the result.

With this tool, we can construct an envelope follower that will track the amplitude of a signal. In Csound, the `rms` opcode estimates the RMS amplitude of a signal and we can use it to shape the envelope of a sound. In this example, we read the data from a file containing a piano note and apply the envelope to a sine wave:

```
gSname = "pianoc2.wav"
instr 1
 k1   = rms(diskin:a(gSname))
 out(oscili(k1,p4))
endin
schedule(1,0,filelen(gSname),500)
```

In Fig. 3.13, we can see how the piano envelope, with a fast attack and a long decay is transferred to the sine wave sound, shaping its amplitude. In this example, it is clear how important this external shape of the waveform is to the dynamic evolution of the sound.

Functions of time such as these can also be generated directly by units called *envelope generators*. These are normally designed to work with piecewise segments of curves, which are concatenated to generate a given shape. Curves come in various forms, but there is a clear distinction between linear and exponential types.

Linear envelopes generate straight lines between points. These are simpler to work with, but not always the most appropriate. Our perception of amplitudes and frequencies is not linear. For instance, we perceive a change from 100 to 200 Hz (a difference of 100) to be very different from 200 to 300 Hz. Likewise, a change of amplitude of say, from 0.25 to 0.5 will not be felt to be the same increase as one from 0.5 to 0.75. We actually judge these in terms of their ratios: so 100:200

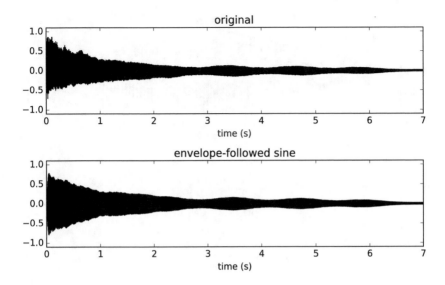

Fig. 3.13 Applying an envelope extracted from a piano tone (top) to a sine wave sound (bottom).

is perceived as the same interval as 400:800; in amplitude, 1:0.5 and 0.5:0.25 are heard as similar amounts of change. For this reason, exponential curves, which are based on ratios rather than differences, are better suited to some applications.

In Csound, linear and exponential envelopes are represented by the `linseg` and `expseg` families of opcodes, respectively. The differences between these two can be appreciated in Fig. 3.14, where we can see these shapes applied to a sine wave.

Multi-segment envelopes can be defined with any number of stages, and Csound has a number of opcodes that allow an arbitrary number of these (in addition to `linseg` and `expseg`), with different characteristics. Alternatively, envelope generators can also use standard three- or four-section models. The classic model is the trapezoid envelope generator `linen`, which has three stages: attack, sustain, and decay. It is usually defined with arguments for the attack and decay times, and the total duration. The sustain period is defined as what is left over of the envelope length, once the attack and decay are accounted for. In addition to this, another very common model is the attack-decay-sustain-release (ADSR) four-stage envelope, which is similar but adds an extra segment between attack and sustain, called decay here (Fig. 3.15).

Envelopes can be applied to all synthesis parameters. It is common to apply, for instance, a pitch inflection to a sound to model certain sounds that either rise or fall in pitch as they develop. It is also very useful to link amplitude changes with timbral modifications, as we will see later. A typical aspect of some instrumental sounds (e.g. brass) is that changes in intensity are also correlatedwith the brightness (the relative weight of higher partials).

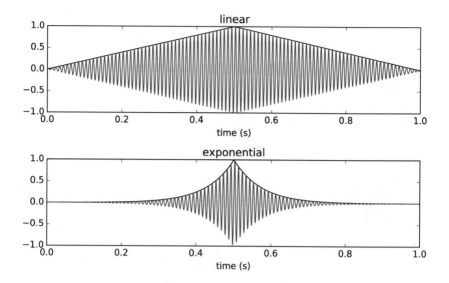

Fig. 3.14 Applying linear and an exponential envelopes to a sine wave.

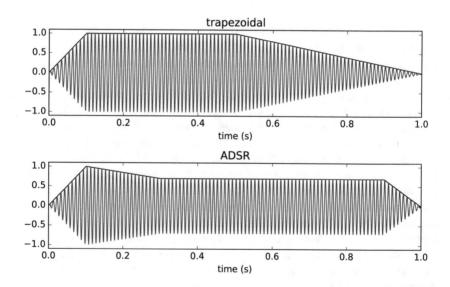

Fig. 3.15 Trapezoidal and ADSR envelopes as generated by the opcodes `linen` and `adsr`.

3.3.2 Modulation

Modulation can be defined as the periodic or aperiodic variation of a parameter within a certain range. It does not have the one-shot nature of an envelope, but this distinction can be slightly blurred in the case of envelopes with a significant number of stages. Modulation can be bipolar, i.e. it modifies a parameter below and above a mean value, or unipolar. The two typical parameters to which modulation is applied are amplitude and frequency, but, again, it is not limited to these.

Given the fact that modulation involves some form of repetition, the concept of rate then emerges in this context. Together with the width, or amount, of modulation, we have the two main parameters that control this process. The nature of the source (periodic, aperiodic, etc.) will also determine the characteristics of the resulting sound.

Any time-varying signal source can be used as a modulator. Typically, as done in the cases considered here, we usesources whose spectrum is mostly composed of low frequencies, below the audio range (< 20 Hz), but more generally, that is also not a limitation[5].

The classic names for amplitude and frequency modulation (AM and FM) in the musical literature are *tremolo* and *vibrato*, respectively, although the two terms sometimes get mixed up. The typical approach to implementing these is to add the modulation source to the parameter value. With AM, we then observe a fluctuation of the amplitude from a minimum to a maximum, which at the extreme will range from 0 to full volume. This can be shown in a simple example with sinusoids:

```
instr 1
 a1 = oscili(p6,p7)
 out(oscili(a1+p4,p5))
endin
schedule(1,0,2,0dbfs/2,400,0dbfs/2,5)
```

where we have 100% modulation. This is because the amplitude A of the modulator is the same as the carrier (the modulated oscillator), and thus at the maximum point of the modulating wave the amplitude of the signal is $2A$ (= 0dbfs), and at the minimum, the amplitude is 0. In Fig. 3.16, we see a plot of the modulated wave with the modulation source superimposed, where we can see that the maxima and minima of the modulation coincide with full amplitude and 0, respectively.

A variation of AM exists where we feed the modulation signal directly to the parameter, without adding it to a mean value. In this case, as the amplitude changes from positive to negative, we see a flipping of the wave shape as all of its values are multiplied by a negative quantity. This is given the name *ring modulation* (RM) and is equivalent to the multiplication of two signals. A simple example with sine waves is shown below:

---

[5] We will see in particular that in the case of frequency modulation, for instance, the use of audio-range modulators leads to a complete new way of synthesising sounds. This will be discussed in detail in Chap. 4.

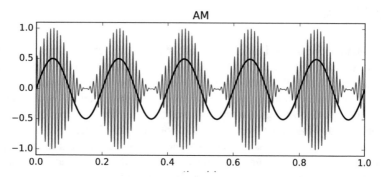

Fig. 3.16 AM wave and its modulation source.

```
instr 1
 a1 = oscili(p6,p7)
 out(oscili(a1,p5))
endin
schedule(1,0,2,0dbfs/2,400,0dbfs/2,5)
```

We can see from the plot in Fig.3.17 that the minima are now reached when the modulator is zero and the maxima when the modulator is at its absolute maximum peak amplitude. Closer inspection will also show that the wave is phase-reversed once at each modulation cycle. The perceived rate of modulation is doubled with regard to the AM example with the same parameters, and the maximum signal amplitude is halved.

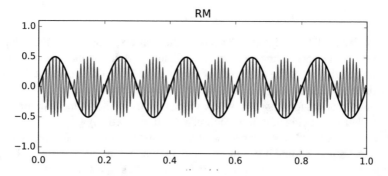

Fig. 3.17 RM wave and its modulation source.

RM starts to become more interesting when both signals are at audio rates. The result will contain a spectrum composed of the sums and differences of all components in the two sources. If we look again at the simple sinusoidal example and raise the modulation frequency to, say, 300 Hz, the result is a wave composed of two components at 100 and 700 Hz (Fig. 3.18). This can become particularly complex if we have sources with many partials.

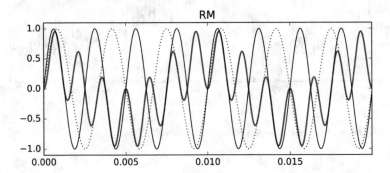

Fig. 3.18 Audio-rate RM, with two modulators (solid and dotted lines), and its output (thick lines).

There is one important application of RM in noise generation that is worth considering. If we employ one of the band-limited noise generators outlined in Sect. 3.2.2, it is possible to shift the spectrum of the noise generator to be centred at a given frequency if we multiply it by a sinusoid. From what we have discussed above, RM will result in the sums and differences of the two spectra. If the noise components are concentrated around 0 Hz, then RM will create a spectrum made up of a band around the sinusoid frequency. This principle is called *heterodyning*.

The following example demonstrates the idea:

```
instr 1
 a1 = oscili(1,p6)
 out(randi(p4,p5)*a1)
endin
schedule(1,0,2,0dbfs,500,1000)
```

where most of the energy in the spectrum is centred on a band 500 Hz wide around 1000 Hz. The `randi` frequency determines the bandwidth and the sinusoid, its centre (Fig. 3.19). Similarly, other band-limited noise generators can be employed to model noise that is centred around a certain part of the spectrum using this approach.

FM, similarly to AM, can be demonstrated with a simple sinusoidal example, whose output is shown in Fig. 3.20:

```
instr 1
 a1 = oscili(p6,p7)
 out(oscili(p4,p5+a1))
endin
schedule(1,0,1,0dbfs/2,100,50,5)
```

In this case, we can see that the frequency of the carrier waveform glides from a maximum that is determined by the modulation width and the original frequency to

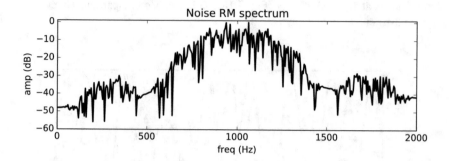

Fig. 3.19 The spectrum of band-limited noise modulated with RM.

a minimum, the difference between these two. In the example above, we have quite a wide modulation. Typical uses in vibrato simulation are more limited.

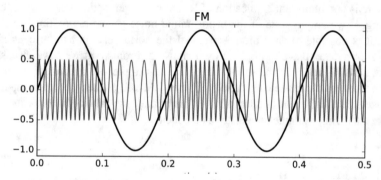

Fig. 3.20 Frequency modulation signal with the (normalised) modulator waveform superimposed (thick line).

Finally, we have seen that non-periodic signals can also be used for modulation in the RM example above. More generally, various types of periodic wave shapes as well as noise generators can be used as low-frequency (LF) sources for parameter control. LF oscillators are often specialised units that are designed to work in the 0 to 20 Hz range. For instance, given that these operate at sub-audio frequencies, it is possible to use some non-band-limited geometric waveforms for these applications (which would have to be avoided in direct audio generation). Sample-and-hold noise generators can also be used to generate random fluctuations for pitch sequences (in a more generalised concept of vibrato). The following example creates a run of pitches between 220 and 660 Hz, with each lasting $\frac{1}{2}$ second:

```
instr 1
 a1 = randh(p6,p7)
```

```
 out(oscili(p4,p5+a1))
endin
schedule(1,0,10,0dbfs/2,440,220,2)
```

Wave shape modulation

Beyond the usual amplitude and frequency parameters, other types of modulation can be done. Particularly interesting are the changes in wave shape that can be achieved in a dynamic way. This, of course, is mainly the role of filters, which we will examine in Sect. 3.4. However, it is possible to modulate the shape of a source waveform by other means. Excluding more explicit distortion, which we will see in Chap. 4, there are two main forms of shape modification: *pulse width modulation* and *hard sync*.

Pulse width modulation (PWM) allows us to manipulate the *duty cycle* of a square waveform. This can be defined as the duration ratio between the positive and negative values of an ideal bipolar square wave. A duty cycle of 50% means that the wave period is divided evenly between its positive and negative values. At 0, the signal is constant (DC) and negative, and at 100%, it is constant and positive. In between these values, we have pulse waves of different widths, with a phase inversion at 50% (Fig. 3.21). The spectrum changes from pulse-like (all harmonics with almost the same amplitude) to square (odd harmonics, weighted by $1/N$, where N is the harmonic number).

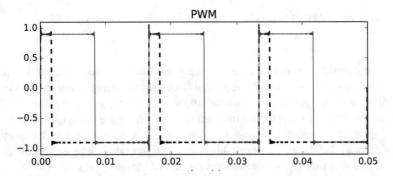

Fig. 3.21 Pulse width modulation with 50% (solid line) and 10% (dashed line) duty cycles.

The following example implements PWM using `vco2`:

```
instr 1
 kpw = p7*(.51+ .49*oscili(1,p6))
 a1 = vco2(p4,p5,2,kpw)
   out(linen(a1,0.01,p3,0.1))
endin
```

```
schedule(1,0,60,0dbfs/2,200,0.5,0.5)
```

Similarly, it is possible to change from a sawtooth wave to a triangle wave, with similar changes from a full harmonic spectrum $(1/N)$ to odd harmonics only $(1/N^2)$. In Csound, `vco2` also implements this type of modulation.

Hard sync is a type of waveform modulation that requires a master clock period to be applied to a slave waveform of a shorter period. This is normally done, in analogue synthesisers, by synchronising two oscillators so that the phase of the slave is reset at the same time as the master. If the slave wave is an integer multiple of the master, then there is no change of shape, otherwise, the slave wave is cut short, re-shaping it. In Fig. 3.22, we see a hard sync waveform whose original frequency was 3.5 times the master frequency. The result is a waveform with a fundamental set to the master, but with a spectrum that includes some dips and resonances.

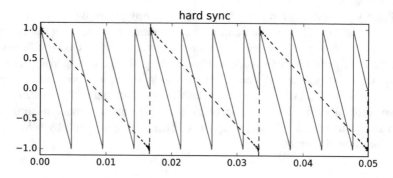

Fig. 3.22 Slave (solid line) and master (dashed line) of hard sync waveform.

By modulating the slave frequency, we get a filtering effect that can be quite dramatic, especially with low fundamental frequencies. To implement this, we need to employ a master phase source whose output is scaled according to a slave:master ratio and then used to read from a function table. As the phase is reset, the waveform is cut short. However, this generates a discontinuity in the output signal, causing a lot of aliasing distortion. To reduce this, we can employ another signal, produced in sync with the master phase that smooths the end of the waveform.

The shape of this signal should be equal to 1 except at the end, where it should be reduced to zero. We can build this with GEN 5, which creates exponential curves. Because these cannot be zero at any point, we create an inverted table such as

```
is = 16384
gil = ftgen(0,0,is,5,0.001,is*0.8,0.001,is*0.2,1)
```

and then subtract the signal from minus 1. The output will then range from 0.999 to 0. With this in place, all we need to do is to multiply the slave waveform by this smoothing signal. The following example implements these ideas, using a sawtooth wave created by `vco2init`:

```
is = 16384
ift = vco2init(1,2)
gi1 = ftgen(0,0,is,5,0.001,is*0.8,0.001,is*0.2,1)

instr 1
 ift = vco2ift(p5,0,0.3)
 aph = phasor(p5)
 krt = oscili(1,0.5)
 asig = tablei:a(aph*(krt+3.5),ift,1,0,1)
 asmt = tablei:a(aph,gi1,1,0,1)
    out(p4*(1-asmt)*asig)
endin
schedule(1,0,10,0dbfs/2,130)
```

Both PWM and hard sync are very common types of modulation, and are widely employed in the digital modelling of analogue synthesis.

3.4 Spectral Modification

We are about to complete our look at the components of the source-filter instrument model (Fig. 3.23). In the context of this design, we have seen some examples of sources and dynamic shapers to control aspects of the sound over time. In this section, we will look at how the spectra of sources, which so far have been presented as static or unchanging, can be modified to suit our desired target signal through the use of filters. It is important to note that such transformations of the source sounds need to be dynamic, and for this reason the model also includes a direct shaping of the filter parameters.

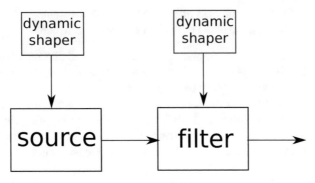

Fig. 3.23 Source-filter instrument model.

Filters are DSP operations that modify the signal in one or more ways. The most typical effect that they have is on the amplitudes (or magnitudes) of the spectrum. They can boost or cut certain frequency regions, depending on the type of filter utilised and the parameters used. In addition, filters can affect the timing of signals, by adding delays to them, sometimes acting at different frequencies, non-linearly, or at times linearly om the whole spectrum.

3.4.1 Filter types

Filters can be characterised in a number of ways. Firstly, we can define them in terms of their effect on the amplitude spectrum. They can also be classified in terms of the way they are constructed, and finally whether they have a linear or non-linear effect on the phase (timing) of signals.

Amplitude spectrum

There are five fundamental types of filters according to their effect on the amplitude spectrum of an input sound (Fig. 3.24):

1. Low pass (LP, or high-cut): filters that preserve lower frequencies, while cutting the higher parts of the spectrum.
2. High pass (HP, or lo-cut): filters that cut low frequencies and let higher components pass.
3. Band pass (BP): a portion of the spectrum within a certain bandwidth centred on a given frequency is passed, while components outside it are suppressed.
4. Band reject (BR): conversely, a filter that cuts frequencies within a certain range of the spectrum.
5. All pass (AP): filters that pass all frequencies with their amplitude unchanged, but with an effect on the phase of the input signal.

The effect of a filter on the amplitude of an input is called the *amplitude response*. The five shapes above indicate the overall curve of the amplitude response of a filter. They determine the pass and stop bands of the filter.

We should also mention the case of *resonant* filters, which in addition to the overall shape (low-, high- or band-pass) have the effect of boosting significantly a small range of frequencies in the spectrum. These filters are particularly interesting from the point of view of musical applications, owing to their unique sound quality.

Delays

The basic principle behind the operation of a digital filter is the use of delays. These are implemented in a computer by storing samples of a signal in memory and re-

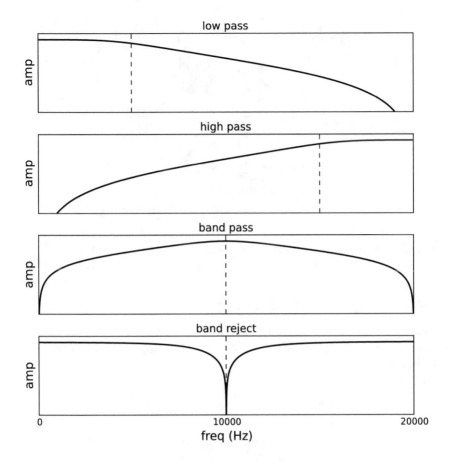

Fig. 3.24 Amplitude response types.

leasing them after a certain amount of time, which can range from one to many sampling periods. The filter output can be a mix of any number of delayed streams. To each one, a gain, called a *coefficient*, is applied. The resulting amplitude response is determined by the combination of delays and coefficients employed.

Filters can be classified by the types of delays they employ. A filter that combines the original signal with delayed input streams is called a *feedforward* filter. A simple example is given by the following Faust code (and the flowchart in Fig. 3.25), with a 1-sample delay and coefficients {0.5, 0.5} :

```
process = _ <: *(0.5) + 0.5*_';
```

where we see the single quote operator (') representing a 1-sample delay.

If, instead, we take the output, delay it and combine it with the input signal, then we have a *feedback filter*:

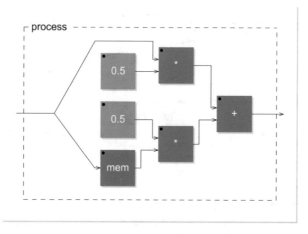

Fig. 3.25 Feedforward filter example, where the *mem* box represents a 1-sample delay.

```
process = _ + *(0.5) ~ *(0.5);
```

where we note that the feedback path implies a 1-sample delay of the output, as shown in Fig. 3.26. Delays of the two types can be combined in filter structures if required.

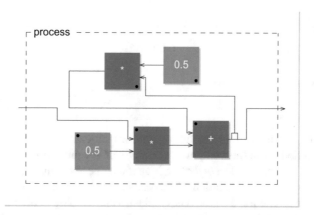

Fig. 3.26 Feedback filter example, with an implied 1-sample delay of the output (indicated by the small box in the feedback path).

Another way of describing these processes is through the concept of the *impulse response*. A signal that is composed of a unit sample followed by zeros is called a unit-sample impulse. If we feed this into a system, we get its impulse response. A filter with feedforward delays will have a finite impulse response: after a certain amount of time, the impulse response is zero, since the unit sample has passed through all the delays into the output. This makes it a finite impulse response (FIR) filter. On the other hand, if the filter includes a feedback stream, then the impulse response will be infinite, making it an infinite impulse response (IIR) filter.

Length

Regardless of the types of delays it uses, we can classify a filter by its delay line length. If it contains 1-sample delays, then it is called a first-order filter. If it contains longer delays, the order is determined by the longest delay. FIR filters tend to be of higher order than IIR filters, as they need more elements to realise a given amplitude response.

In the case of IIR filters, we associate the order of the filter with the stop-band rolloff, or how steep the amplitude response curve is outside the pass band. In general, we have a 6 dB per octave rolloff in a first-order filter. This means that the amplitude drops by $\frac{1}{2}$ every octave (a 6 dB drop is roughly equivalent to an 1:2 ratio). Consequently, 2nd-order filters exhibit a 12 dB/octave rolloff, and 4th-order, a 24 dB/octave rolloff. In general, the higher the order, the more selective the filter will be.

Phase spectrum

Every filter will delay a signal in a certain way, which effectively means that it modifies the phase of an input. The action of a filter over the whole phase spectrum is called its *phase response*. This curve can be linear, where effectively the same time delay is applied to all frequencies, or it can be non-linear, where different parts of the spectrum are delayed by different amounts. FIR filters can be designed to have a linear phase response, which is their main advantage over IIR filters, which are non-linear in this respect. All-pass filters can be used to exploit this aspect to modify the phase of signals in some desired way without affecting their amplitude.

3.4.2 IIR versus FIR

IIR filters tend to be used more widely than FIR ones for source-filter instrument models because they are more flexible, simple to apply, and can work more easily with time-varying parameters. On the downside, they are not linear in phase, but that tends to be less of an issue in this application. FIR filters are often designed by defining a fixed amplitude response shape from which the coefficients are calculated. IIR filters, on the other hand, come in pre-packaged designs, with some basic fixed characteristics (e.g. overall shape and order) that depend on one or two parameters to determine the final amplitude response. These can be varied dynamically by envelopes or modulators.

IIR filter parameters generally determine how the various types of curves (LP, HP, BP, BR, or AP) will actually work on the spectrum. These are:

1. Frequency: this controls where the cut-off point is, for low-pass and high-pass filter, or the centre frequency in the case of band-pass or reject filters.
2. Bandwidth: this determines the width of the pass or reject bands, or of the resonance region.

Both frequency and bandwidth are measured in Hz. A related parameter that is often used to control how filters behave is called the Q, defined as

$$Q = \frac{f}{B} \tag{3.6}$$

for a frequency f and bandwidth B. A Q control is sometimes useful as it defines a given interval of frequencies based on a ratio, rather than a linear difference (given by the bandwidth). As we have already discussed, this might be more musically meaningful given that we tend to perceive pitch on the basis of ratios of frequencies.

IIR filters can be divided into a number of different design families. Four of these are of note: tone, resonators, Butterworth, and low-pass resonant filters.

Tone

Tone filters come in low- and high-pass forms. They have a 6 dB/octave amplitude response rolloff, as shown in Fig. 3.27. These simple filters are useful for a gentle removal of high or low frequencies, without too much attenuation. In Csound, the low-pass version of this filter is implemented in the tone opcode[6].

A flowchart for this type of filter is shown in Fig. 3.28. The Faust program for the tone filter is:

```
import("math.lib");
fc = hslider("freq",0,0,20000,1);
cost = cos(2*PI*fc/SR);
c = 2 - cost  - (sqrt((2 - cost)^2) - 1);
process = _ + *(1.-c) ~ *(c);
```

The high-pass filter can be obtained by replacing the c coefficient calculation by this:

```
c =  (sqrt((2 + cost)^2) - 1) - 2 - cost
```

Tone low-pass filters can also be used very effectively to smooth control signals, removing discontinuities in them. They are also used in the process of computing

---

[6] Its companion high-pass opcode, atone is implemented slightly differently from the one shown here.

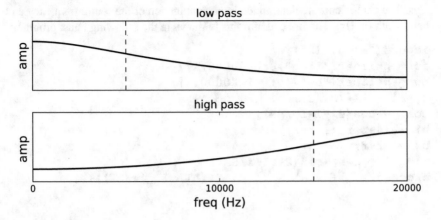

Fig. 3.27 Low-pass and high-pass tone filter amplitude responses, cutoff frequencies indicated by the vertical lines, amplitude range 5 to −60 dB.

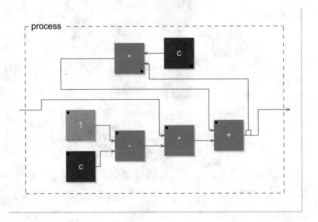

Fig. 3.28 Tone filter flowchart.

the RMS value of a stream to calculate an average of the signal samples. In both cases, the cut-off frequency is typically low, under 10 Hz.

Resonators

Resonators are 2nd-order band-pass filters that can be used in a variety of applications. The overall design is shown in Fig. 3.29, where we can see that the filter is a combination of the input signal (scaled) and one- and two-sample feedback delays. The coefficients are determined by a combination of the centre frequency and bandwidth (in Hz). The full code for this is shown in the following Faust program:

```
import("math.lib");
fc = hslider("freq",0,0,20000,1);
bw = hslider("bw",10,0.1,20000,1);
r = 1 - PI*(bw/SR)
cost = cos(2*PI*fc/SR);
b1 = (4*r*r/(1+r*r))*cost;
b2 = -r*r;
a = (1 - r*r)*sin(2*PI*f/SR);
process = _ + *(a) ~ (_ <: (*(b1) + _'*(b2))) ;
```

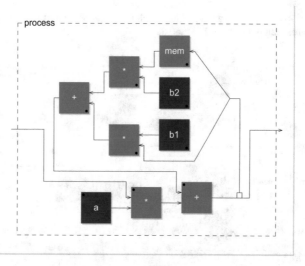

Fig. 3.29 Basic resonator design.

This is implemented in Csound as the `reson` opcode. There are two variations of the code that include a feedforward section, defined by the following code (Fig. 3.30):

```
process = _ + (_ <: *(a1) + _''*a2)
              ~ (_ <: (*(b1) + b2*_')));
```

where the double-quote operator (`''`) is used to indicate a two-sample delay.

In Csound, we have two variations of the filter based on this design, called `resonr` and `resonz`, with slight differences in the way the two-sample feedforward delay coefficient is set. The amplitude response for the three filters in this family is shown in Fig. 3.31, where we can see the differences, which are mainly to do with how the curve behaves at 0 Hz and at the Nyquist frequency.

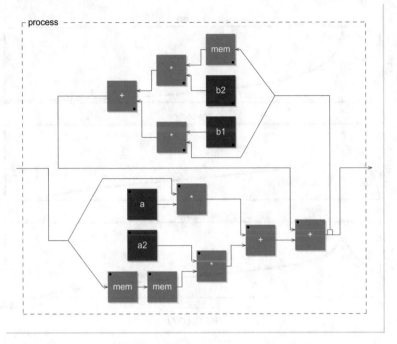

Fig. 3.30 Resonator design variation as implemented by `resonz` and `resonr`.

Butterworth

There are a number of different digital filter designs that are derived from classic analogue filters. The most common of these are the Butterworth design, which offer a maximally flat pass band and good stop band attenuation. They can be constructed to have the BP, LP, HP, and BR amplitude responses plotted in Fig. 3.24.

These filters are typically presented in 2nd-order sections composed of one- and two-sample feedforward and feedback delays (Fig. 3.32), to which we supply different coefficients according to the required amplitude response. The following example shows the code for a Butterworth band-reject filter:

```
import("math.lib");
fc = hslider("freq",0,0,20000,1);
bw = hslider("bw",10,0.1,20000,1);
r = 1 / tan(PI*bw/SR);
cost = cos(2*PI*fc/SR);
a = 1/(1 + r);
a1 = -a*cost;
a2 = a;
b1 = -a1;
b2 = -a*(1 - r);
```

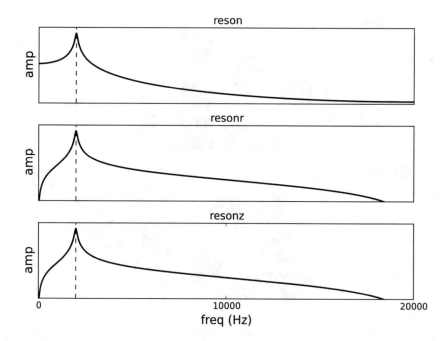

Fig. 3.31 Amplitude responses for three variations ot the resonator design; centre frequencies indicated by the vertical lines, amplitude range 5 to −60 dB.

```
process = _ + (_ <: *(a0) + (a1*_' <: (_ + a2*_')))
              ~ (_ <: (*(b1) + b2*_'));
```

To get the other response curves, we need to calculate the five coefficients $(a_0, a_1, a_2, b_1, b_2)$ using different formulae (see [44] for more details). In Csound, we have four opcodes, `butterlp`, `butterhp`, `butterbp`, and `butterbr`, that implement these filters .

Low-pass resonant

Resonant filters are very common in musical applications. These are characterised by a low- or high-pass curve that includes a peak at the cutoff frequency. One of the most famous of the analogue filters that implement this type of amplitude response is the 4th-order Moog ladder filter, from which a number of digital implementations have been made. In general the design of these filters, especially if derived from analogue models, can be quite complex. However, we can give a general idea here. The 4th-order Moog-like designs are constructed out of four 1st-order IIR low-pass sections. They also include a feedback path from the output of the filter back into its

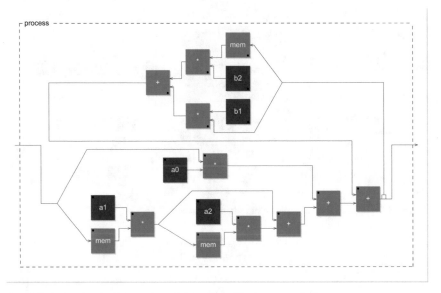

Fig. 3.32 2nd-order IIR filter section of the type used by Butterworth filters.

input, which provides the resonance effect. The following Faust code excerpt gives
a general outline of the filter structure:

```
process = (*(r) + _ :
        _  + *(w) ~ (_ <: _ - w*_) :
        _  + *(w) ~ (_ <: _ - w*_) :
        _  + *(w) ~ (_ <: _ - w*_) :
        _  + *(w) ~ (_ <: _ - w*_))
        ~ (_ <: _ + _': *(0.5)) ;
```

The parameters w and r are calculated from the cutoff frequency and the resonance, respectively. The flowchart for this design is shown in Fig. 3.33. Missing
from this simple schematic are the non-linear components, which are applied to the
feedback paths to distort slightly the signal. These make the filter more realistic, but
also introduce foldover, which requires further techniques such as oversampling to
be employed for aliasing suppression.

In Csound, the opcodes `moogvcf`, and `moogladder` and the `mvclpf` family
are examples of different designs based on the original Moog low-pass filter. In Fig.
3.34, we have plots of three amplitude responses from the `moogladder` opcode,
using different amounts of resonance (whose normal range is from 0 to 1).

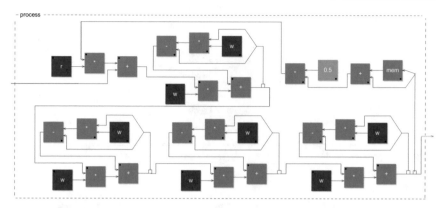

Fig. 3.33 24 dB/octave resonant low-pass filter structure.

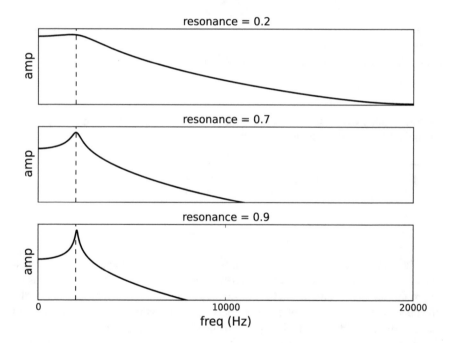

Fig. 3.34 Amplitude responses for a 4th-order low-pass resonant filter (Moog ladder design) with different amounts of resonance.

3.4.3 Multiple filters

Filters can be connected together in two types of arrangements: in series (sequentially) or in parallel.

Series connection

Filters connected in series are working on the already modified outputs of its predecessors in the chain. This means that their combination can be thought of as one single, longer filter. The total amplitude response is the product of all the individual amplitude responses of the filters in the chain. A common application fof this type of arrangement is to make a higher-order filter out of smaller sections. For instance we can make a 24 dB/octave filter based on a Butterworth design by placing two such 2nd-order sections in series and applying the same parameters to each of them. This is shown in Fig. 3.35, where we plot the output of one-, two-, three-, and four-section Butterworth BP filters.

Parallel connection

Parallel connections of filters are constructed by feeding the same input to a bank of different filters and mixing their output. In this case, the overall response of the filter will be a sum of the original responses and the total order of the connection will be that of the longest filter. As far as the amplitude response of a parallel connection of band-pass filters is concerned, the individual responses will contribute to creating a series of separate peaks if their centre frequencies are coincident (Fig. 3.36).

3.5 Instrument Examples

In this section, we will apply some of the ideas developed in the first sections of this chapter. We will show how the source-filter model of signals can be used to create a variety of instruments for music and sound design[7].

3.5.1 Multiple detuned sources to resonant filter

The first design we will study is the use of several oscillators, whose fundamentals diverge a little, feeding into a resonant filter, with shaping of parameters with differ-

---

[7] All complete Csound examples in this and other sections of the book can be hosted in Python with similar code to Listings 2.1 and 2.2.

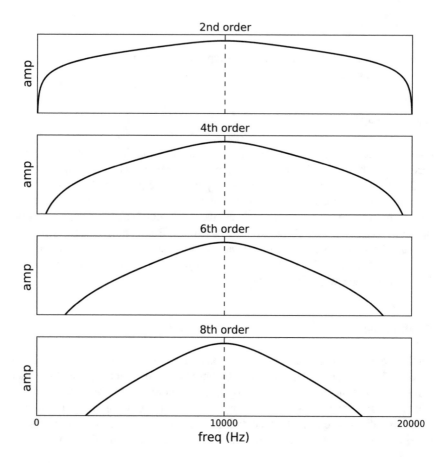

Fig. 3.35 Amplitude responses for 2nd to 8th-order Butterworth band-pass filters, constructed out of one to four second-order sections.

ent envelopes. We will choose vco2, which as we have seen, provides band-limited source waveforms with various shapes. The idea behind using a mix of detuned oscillators is that this allows for a thickening of the source signal, which also injects some liveliness into the static waveform signal. This is akin to the *chorus* effect that arises from multiple instruments playing together. We can choose any number of oscillators; in this example, three will be used.

The detuning and the waveform shapes can be controlled through instrument parameters. There are three vco2 modes to select the basic sawtooth, square, and triangle waves (0, 10, 12), and we use an array to simplify the selection:

```
imodes[] fillarray 0,10,12
imd = imodes[p6]
idt = p8
```

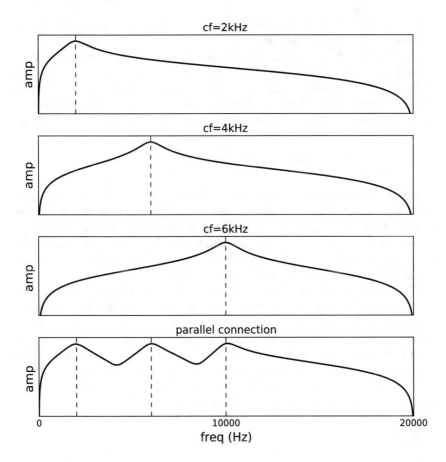

Fig. 3.36 Parallel connection of three Butterworth BP filters centred at 2, 6, and 10 kHz, all with the same bandwidth.

```
a1   = vco2(p4, p5,imd)
a2   = vco2(p4, p5*(1+idt),imd)
a3   = vco2(p4, p5*(1-idt),imd)
amix = a1 + a2 + a3
```

The mix of signals is then put into a `moogladder` filter, whose frequency is shaped by an envelope. We get the resonance from an instrument parameter:

```
kcf  = expsegr(p5*2,0.01,p5*10,p3-0.01,p5,0.5,p5)
afil = moogladder(amix,kcf,p7)
```

Note the use of `expesegr`: this is one of the envelopes with an added *release* period, which is used after the instrument is turned off. The final two arguments define the release time and the final target value at the end of it. We use the fundamental (p5) as the basis for calculating the filter cutoff frequency, so higher pitches will sound brighter than lower ones.

In a similar way, we can use another one of these 'r' envelope generators to shape the amplitude of the sound. In this case, we use `linenr`, which is a trapezoid attack-sustain-decay whose final stage is only started after turnoff. After the attack, the note is sustained until release. After this, the decay in this example takes 0.5 secs. The final parameter of `linenr` determines the rate of decay, since in this case, we have an exponential curve in this stage.

```
aenv = linenr:a(afil,0.01,0.5,0.01)
   out(aenv)
```

This instrument can be used in a variety of settings. In the full example, we provide a sequence of 1000 events (notes), with varying parameters taken from random generators. This is built using a `while` loop construction, which iterates a block of code between the words do and od until the condition becomes false. We use a counting variable (`icnt`) to keep track of the loop count, which is used in the condition check and to calculate the start time of events.

The `gauss` opcode, which we have seen in a different guise earlier in this chapter, is used to change parameters around a certain centre (the Gaussian distribution tends towards the mean, which is zero in this case). The `rnd` is just a simple unipolar pseudo-random generator. Fundamental frequencies are obtained from pitches given in an array (`inot[]`). These are written in octave.pitch-class format, the number before the decimal point indicates the octave (e.g. 8 is the middle octave of the piano) and the two decimal values after that define the 12-tone equal-temperament pitch class (00 = C, 01 = D♭, ..., 11 = B, see Sect. 1.2.1). The function `cpspch()` converts these to cps (Hz) frequency using the 12 TET scale:

Listing 3.10 Multiple oscillator + resonant filter example.

```
instr 1
 imodes[] fillarray 0,10,12
 imd = imodes[p6]
 idt = p8
 a1 = vco2(p4, p5,imd)
 a2 = vco2(p4, p5*(1+idt),imd)
 a3 = vco2(p4, p5*(1-idt),imd)
 amix = a1 + a2 + a3
 kcf = expsegr(p5*2,0.01,p5*10,p3-0.01,p5,0.5,p5)
 afil = moogladder(amix,kcf,p7)
 aenv = linenr:a(afil,0.01,0.5,0.01)
    out(aenv)
endin

icnt = 0
ibpm = 120
inot[] fillarray 6.00,8.03,7.07,8.10,9.02
while icnt < 1000 do
```

```
itempo = 60/ibpm
istart = icnt*itempo*0.25
idur = 0.4 + gauss(0.3)
iamp = 0dbfs/6 + gauss(0dbfs/9)
ifreq = cpspch(inot[abs(int(gauss(icnt/10)))]%5])
ires = 0.5 + gauss(0.4)
iwave = int(rnd(3))
idetune = 0.005 + gauss(0.004)
schedule(1,istart,idur,
         iamp,ifreq,
         iwave,ires,
         idetune)
 icnt += 1
od
```

This instrument can be slightly modified to take in other parameters, such as attack and decay time, and cutoff frequency scaling. It can also be employed in many different ways to the one in the example above, which has emphasised its use for short, rhythmic, sounds.

3.5.2 Synthetic vocals using filter banks

The source-filter model can be applied very effectively to the synthesis of voices. It is possible to decompose the vocal mechanism into a sound-producing element, which is concentrated mainly in the glottis, and a set of modifiers, made up of all the complex moving parts, mostly situated in the head and mouth.

In this example, we will be looking to generate vowel sounds, which are sometimes called *voiced* signals in the speech-processing literature, indicating their · pitched nature. The basic principles are again the use of a suitable periodic waveform as a source and an adjustable filter capable of multiple resonance peaks. These are used to model the spectral contours that characterise each vowel. In Fig. 3.37, a plot is shown of the amplitude response of some filters that can be used to model different vowels as sung by a low-range female voice.

In order to construct filters that are capable of this task, we need to put together a parallel set of five band-pass components, each one targeting a specific region of the spectrum. To simulate a vowel, we need to know the centre frequency, bandwidth and relative amplitude of each peak. Conventionally, we parametrise the vocal spectrum in three to five of these resonance areas, known as *formants*. For this example we will use a filter bank composed of five band-pass elements.

To allow us to try 2nd and 4th-order band-pass components of various types, we use a Csound code structure called a *user-defined opcode*, or UDO. This allows us to package an element of the instrument separately and then use it as another opcode in the language. The definition must list any output and input parameters that the UDO will need. In this particular case, we define a `Filter` with the same usual

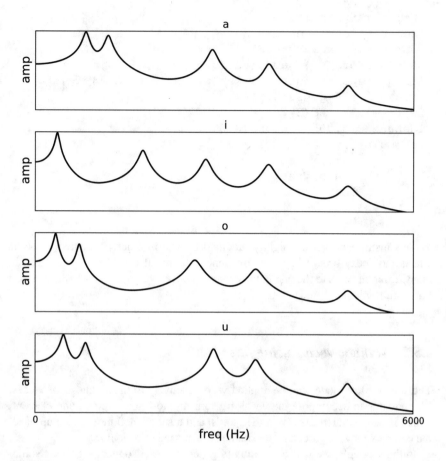

Fig. 3.37 Four filter amplitude responses modelling different vowels as sung by a low-range female voice.

parameters as any band-pass filter (input, frequency, and bandwidth), and make it a 4th-order filter with two resonators:

```
opcode Filter,a,akk
 as,kf,kbw xin
 as = reson(as,kf,kbw,1)
 as = reson(as,kf,kbw,1)
    xout(as)
endop
```

To make it a 2nd-order filter, we can just remove (or comment out) one of the lines. We can also try other band-pass models (e.g. a Butterworth filter), which will just replace the resonator. With this in place, we have our filter bank:

```
af1 = Filter(a1*ka0,kf0,kb0)
af2 = Filter(a1*ka1,kf1,kb1)
af3 = Filter(a1*ka2,kf2,kb2)
af4 = Filter(a1*ka3,kf3,kb3)
af5 = Filter(a1*ka4,kf4,kb4)
```

Each filter has its own input scaling variable, frequency, and bandwidth, and is fed the same source signal. For this, we have a choice of waveforms. It is standard to use a pulse wave, but if this is perceived to be too bright, we can opt for a sawtooth wave or even insert a tone filter before the spectral shaping to gently remove some of the high-frequency energy. In this example, we will use a filtered pulse waveform, although this can easily be replaced by another source. Also, to make the signal more realistic, we add some random jitter and vibrato to the fundamental frequency.

We also need to package the frequencies, bandwidths and amplitudes that characterise our vowels. We do this in three matrices stored in k-variable two-dimensional arrays. Each row contains parameters for five formants characterising one of five vowels (a, e, i, o, and u)[8]. In Csound, the `fillarray` opcode can be used to fill an array in row order, and the usual square brackets can be used to access each member (e.g. `ka[0][0]`):

```
kf[][] fillarray 800,1150,2800,3700,4950,
                 450,800,2830,3500,4950,
                 400,1600,2700,3500,4950,
                 350,1700,2700,3700,4950,
                 325,700,2530,3500,4950
kb[][] fillarray 80,90,120,130,140,
                 70,80,100,130,135,
                 60,80,120,150,200,
                 50,100,120,150,200,
                 50,60,170,180,200
ka[][] fillarray 0,4,20,36,60,
                 0,9,16,28,55,
                 0,24,30,35,60,
                 0,20,30,36,60,
                 0,12,30,40,64
```

This allows us to access each vowel by selecting a row of each matrix. In the complete example, we modulate this lookup with a low-frequency noise generator, getting various sequences of vowels. To smooth the transitions, we employ a `port` opcode, which is a tone-like low-pass filter with a half-time parameter that allows the parameter values to glide from one to another. We also make sure that the lowest filter frequency is never lower than the fundamental:

Listing 3.11 Synthetic voices example.

```
opcode Filter,a,akk
```

---

[8] This particular set models a low-range female voice.

```
 as,kf,kbw xin
 as = reson(as,kf,kbw,1)
 as = reson(as,kf,kbw,1)
    xout(as)
endop

instr 1
kf[][] init 5,5
kb[][] init 5,5
ka[][] init 5,5

kf[][] fillarray 800,1150,2800,3700,4950,
                 450,800,2830,3500,4950,
                 400,1600,2700,3500,4950,
                 350,1700,2700,3700,4950,
                 325,700,2530,3500,4950
kb[][] fillarray 80,90,120,130,140,
                 70,80,100,130,135,
                 60,80,120,150,200,
                 50,100,120,150,200,
                 50,60,170,180,200
ka[][] fillarray 0,4,20,36,60,
                 0,9,16,28,55,
                 0,24,30,35,60,
                 0,20,30,36,60,
                 0,12,30,40,64

kv = randh(10,1)
kv = abs(int(kv)%4)
itim = 0.03

kf0 = port(kf[kv][0],itim)
kf1 = port(kf[kv][1],itim)
kf2 = port(kf[kv][2],itim)
kf3 = port(kf[kv][3],itim)
kf4 = port(kf[kv][4],itim)

kb0 = port(kb[kv][0],itim)
kb1 = port(kb[kv][1],itim)
kb2 = port(kb[kv][2],itim)
kb3 = port(kb[kv][3],itim)
kb4 = port(kb[kv][4],itim)

ka0 = port(ampdb(-ka[kv][0]),itim)
ka1 = port(ampdb(-ka[kv][1]),itim)
```

```
ka2 = port(ampdb(-ka[kv][2]),itim)
ka3 = port(ampdb(-ka[kv][3]),itim)
ka4 = port(ampdb(-ka[kv][4]),itim)

kjit = randi(p5*0.03,15,2)
kvib = oscili(p5*(0.03+
            gauss:k(0.005)),
            4+gauss:k(0.05))

kfun = p5+kjit+kvib
a1 = vco2(1,kfun,6)
a1 tone a1,kf0
kf0 = kfun > kf0 ? kfun : kf0
af1 = Filter(a1*ka0,kf0,kb0)
af2 = Filter(a1*ka1,kf1,kb1)
af3 = Filter(a1*ka2,kf2,kb2)
af4 = Filter(a1*ka3,kf3,kb3)
af5 = Filter(a1*ka4,kf4,kb4)

    amix = af1+af2+af3+af4+af5
    out(linen(amix*p4,0.2,p3,1))

endin
schedule(1,0.1,30,0dbfs,cpspch(8.04))
```

In this example, we are driving the instrument with a single voice, but more inventive ways of running the instrument could be employed. For instance, a choral-like effect could be created by starting several voices with a slightly detuned fundamental:

```
ifun = cpspch(8.02)
ivoices = 24
icnt = 0
while icnt < ivoices do
  schedule(1,0.1,30,
          0dbfs/ivoices,
          ifun+gauss(ifun*0.05))
  icnt += 1
od
```

Finally, as we have already hinted, we can adjust the instrument in various ways. Firstly, different filter types can be tried in the UDO, and we can also decrease or increase the number of sections used. We can replace our source by a different generator. Instead of a periodic waveform generator, we could use a noise source. We could remove the vowel modulation and base each sound on a single vowel or a set movement between two or more vowels. Other formant sets could be used to model different types of voices. A variation on this instrument is presented in Appendix B (Listing B.2), where the code simulates a vocal quartet.

3.6 Conclusions

The principles that underlie source-model synthesis are based on the way many real-world instruments work. The model of a sound generator followed by some modification to shape the spectrum is very useful and malleable. We have shown in this chapter a number of ways in which we can describe and design components for this technique.

In particular, we should note two key ideas in all of the examples we studied. The first is that parameter shaping is essential, as almost all interesting sounds have dynamically-evolving spectra. The other is that there are many different ways in which we can create source sounds for further transformations. These points will be very important when we examine other methods of synthesis in the following chapters.

Chapter 4
Closed-Form Summation Formulae

Abstract Closed-form summation formulae provide a compact means of describing different types of spectra. By combining a limited number of sinusoids, we can generate complex waveforms with many harmonic or inharmonic partials. In this chapter, we will look at various formulae, and their implementation and application, from band-limited pulse oscillators to frequency modulation and non-linear distortion synthesis.

In this chapter, we will look at combining a few sinusoidal signals together in such a way that we can obtain a complex spectrum with harmonic or inharmonic partials. These methods are collectively called closed-form summation formulae, as they provide compact solutions for expressions that involve a mix of many sinusoids. They take advantage of well-known relationships that represent the relevant arithmetic series, such as the Fourier series, in closed form. Such methods are also known as *distortion* synthesis, as they involve some sort of non-linear distortion of a sinusoidal signal.

For each technique studied, we will provide an implementation in Python. While this code can be used directly in synthesis applications, it is mostly designed for study and research purposes. In addition to these examples, Csound code implementing these techniques is offered in Appendix B, in the form of UDOs, which can be used in more practical applications.

4.1 Band-Limited Pulse

In Chap. 3, we saw that one of the cases of the Fourier series, which produces a pulse wave, is given by a mix of harmonics with the same weight [105]:

$$\frac{1}{N} \sum_{k=1}^{N} \cos(k\omega_0) \tag{4.1}$$

© Springer International Publishing AG 2017
V. Lazzarini, *Computer Music Instruments*,
https://doi.org/10.1007/978-3-319-63504-0_4

where $\omega_0 = 2\pi f t^{1}$.

It turns out that this can be synthesised by a well-known arithmetic series in closed-form given by

$$\sum_{k=-N}^{N} r^k = r^{-N}\frac{1 - r^{2N+1}}{1 - r} \tag{4.2}$$

Applying this to the Fourier series case, we have [21]

$$\frac{1}{2N} \times \left[\frac{\sin((2N+1)\frac{\omega_0}{2})}{\sin(\frac{\omega_0}{2})} - 1 \right] = \frac{1}{N}\sum_{k=1}^{N} \cos(k\omega_0) \tag{4.3}$$

where $\omega_0 = 2\pi f_0 t$, and f_0 the fundamental frequency in Hz. The parameter N guarantees this to be band-limited, as it dictates the number of harmonics.

When programming this, we need to protect the code against a possible zero in the denominator. In this case, we can substitute 1 for the whole expression, although a definitive solution to the problem might require more complicated logic. An example from first principles written in Python is shown below:

Listing 4.1 Band-limited pulse generator.

```python
import pylab as pl

sr = 44100.
fr = 100.
N = int(sr/(fr*2))

t = pl.arange(0,sr)
s = pl.zeros(sr)
w = 2*pl.pi*fr/sr

for n in t:
    den = pl.sin(w*n/2.)
    if abs(den) > 0.0002:
        s[n] = (1./(2*N))*(pl.sin((2*N+1)*w*n/2.)/den - 1.)
    else: s[n] = 1.

pl.figure(figsize=(8,3))
pl.plot(t[0:440]/sr,s[220:660])
pl.xlabel("time (s)")
pl.tight_layout()
pl.show()
```

---

[1] Note that in many formulae in this chapter, we will be tucking away the time variable t together with the frequency f into the phase angle ω (or θ, etc.). This allows a more unencumbered notation for the various closed forms we will be using. When these are translated into programs, we will still have to be careful about calculating the phase increments correctly, as we saw in Chap. 3, if/when the frequency changes over time.

We need to run the code in a loop, so we can check every sample to avoid a possible division by zero. The code can be modified to provide dynamic spectra (by changing the value of N). In this case we can emulate the effect of a low-pass filter with variable cut-off frequency. This algorithm is implemented in Csound by the `buzz` opcode (see also Listing B.3). It is a better alternative to the pulse waveform mode in `vco2` because it produces a normalised output. The plot generated by the code in Listing 4.1 is shown in Fig. 4.1.

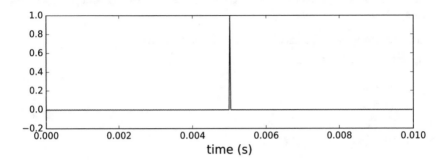

Fig. 4.1 Plot of a band-limited pulse waveform as generated by the program in listing 4.1.

4.2 Generalised Summation Formulae

The principle used in the band-limited pulse algorithm can be expanded to other closed-form summation formulae [70]. This allows a more flexible approach, providing a way to control spectral roll-off, as well as a means of generating inharmonic partials. For band-limited signals, we have

$$\frac{\sin(\omega) - \sin(\omega - \theta) - a^{N+1}(\sin(\omega + (N+1)\theta) - a\sin(\omega + N\theta))}{1 - 2a\cos(\theta) + a^2}$$
$$= \sum_{k=1}^{N} a^k \sin(\omega + k\theta) \tag{4.4}$$

where $\omega = 2\pi f_1 t$ and $\theta = 2\pi f_2 t$. The frequencies f_1 and f_2 (in Hz) are independent. As before, N is the number of components and a is an independent parameter controlling the amplitudes of the partials.

The parameters a and N can be dynamically manipulated to control the spectral roll-off and bandwidth, respectively. Varying these parameters in time allows the emulation of a low-pass filter behaviour. The ω to θ ($f_1 : f_2$) ratio determines the type of spectrum that is generated, which can be either harmonic or inharmonic. To complete the formula, we also require a scaling expression, to avoid the output overflowing, as this method produces a signal whose gain varies with a and N:

$$\sqrt{\frac{1-a^2}{1-a^{2N+2}}} \tag{4.5}$$

The following Python example demonstrate this algorithm, which produces components up to the Nyquist frequency by setting N accordingly (Listing 4.2, see also Listing B.4).

Listing 4.2 Generalised band-limited summation-formula synthesis.

```
import pylab as pl

sr = 44100.
f1 = 100.
f2 = 200.
N = int((sr/2 - f1)/f2)

t = pl.arange(0,sr)
w = 2*pl.pi*f1*t/sr
o = 2*pl.pi*f2*t/sr
a = 0.4

sinw = pl.sin(w)
sinwmo = pl.sin((w-o))
sinwn1o = pl.sin((w+(N+1)*o))
sinwno =  pl.sin((w+N*o))
den = 1.- 2*a*pl.cos(o)  + a*a
scal = pl.sqrt((1.-a*a)/(1- a**(2.*N+2)))
s = scal*(sinw - sinwmo - a**(N+1)*
          (sinwn1o - a*sinwno))/den
```

In Fig. 4.2, we can see how setting the frequency ratio to 1:2 produces a square wave. If the ratio is composed of large or irrational values, then the spectrum will be inharmonic.

With a careful definition of the spectral roll-off, a much simpler non-band-limited expression is possible:

$$\frac{\sin(\omega) - a\sin(\omega - \theta)}{1 - 2a\cos(\theta) + a^2} = \sum_{k=1}^{\infty} a^k \sin(\omega + k\theta) \tag{4.6}$$

where there is no direct bandwidth control. However, it is possible, for instance, to set the -60 dB bandwidth by selecting an a value that will satisfy $a^k > 0.001$ for a given partial k. We can use this to set the maximum value of a, so that any aliased components will be at least -60 dB down from the loudest partial:

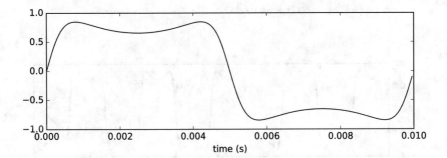

Fig. 4.2 The plot of a band-limited square waveform as generated by the program in listing 4.2.

$$a = 10^{-\frac{3}{N}} \qquad (4.7)$$

where N is the highest component below the Nyquist frequency.

The scaling expression is also simpler:

$$\sqrt{1 - a^2} \qquad (4.8)$$

The code to match this expression is presented in Listing 4.3. A plot of the waveform and spectrum generated by it is shown in Fig. 4.3[2] (see also Listing B.5).

Listing 4.3 Non-band-limited generalised summation formulae synthesis.

```python
import pylab as pl

sr = 44100.
f1 = 500
f2 = 1000.
N = int((sr/2 - f1)/f2)

t = pl.arange(0,sr)
w = 2*pl.pi*f1*t/sr
o = 2*pl.pi*f2*t/sr
a = 0.4
if a > 10**(-3./N):
    a = 10**(-3./N)

sinw = pl.sin(w)
sinwmo = pl.sin((w-o))
den = 1.- 2*a*pl.cos(o) + a*a
scal = pl.sqrt(1. - a*a)
```

---

[2] The code used for plotting this and other figures in this chapter is given in Appendix B, in Listing B.14.

```
s = scal*(sinw - sinwmo)/den
```

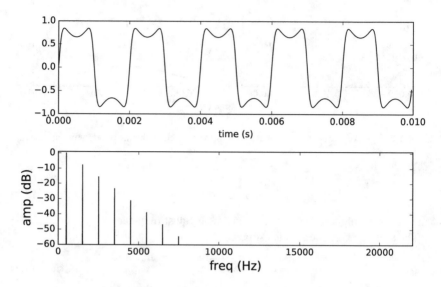

Fig. 4.3 Plot of a square waveform and its spectrum as generated by the program in listing 4.3.

The formulae provided above produce what are called *single-sided* spectra, where the new components are all above ω. They have double-sided equivalents, where the new components are located both above and below a centre frequency. For band-limited spectra, we have

$$
\frac{\sin(\omega)(1 - a^2 - 2a^{N+1}(\cos((N+1)\theta) - a\cos(N\theta)))}{1 - 2a\cos(\theta) + a^2}
$$
$$
= \sin(\omega) + \sum_{k=1}^{N} a^k(\sin(\omega + k\theta) + \sin(\omega - k\theta))
$$
(4.9)

with the scaling set to

$$
\sqrt{\frac{1 - a^2}{1 + a^2 - 2a^{2N+2}}}
$$
(4.10)

A Python program implementing this formula is shown in Listing 4.4. It creates a spectrum that is centred around ω, with frequencies each side of it (also known as *sidebands*; see Sect. 4.3). A plot of the program output is presented in Fig. 4.4.

Listing 4.4 Band-limited generalised summation-formulae synthesis, double-sided spectrum.

```
import pylab as pl
```

```
sr = 44100.
f1 = 3000
f2 = 200.
N = 10

t = pl.arange(0,sr)
w = 2*pl.pi*f1*t/sr
o = 2*pl.pi*f2*t/sr
a = 0.5

sinw = pl.sin(w)
cosmo = pl.cos((N+1)*o)
cosno = pl.cos(N*o)
den = 1.- 2*a*pl.cos(o) + a*a
scal = pl.sqrt(1. - a*a/ (1+a*a-2*a**(2*N+2)))
s = sinw*(1 - a*a - (2*a**(N+1))*(cosmo - a*cosno))/den
s *= scal
```

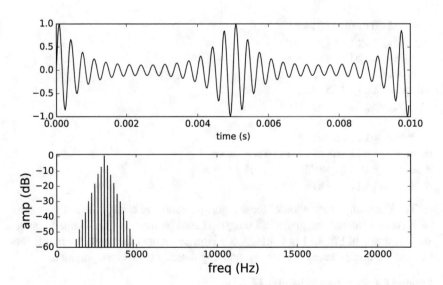

Fig. 4.4 The plot of a waveform and its spectrum as generated by the program in listing 4.4.

The non-band-limited version of Eq. 4.9 is:

$$\frac{\sin(\omega)(1-a^2)}{1-2a\cos(\theta)+a^2}$$

$$= sin(\omega) + \sum_{k=1}^{\infty} a^k(\sin(\omega+k\theta)+\sin(\omega-k\theta)) \tag{4.11}$$

In this case, the scaling is adjusted to

$$\sqrt{\frac{1-a^2}{1+a^2}} \tag{4.12}$$

The program for this formula is given in Listing 4.5, and the waveform and spectrum are shown in Fig. 4.5. As can be seen, the result is identical to that in the band-limited case, which indicates that it is possible, with care, to use this much simpler formulation instead.

Listing 4.5 Non-band-limited generalised summation-formulae synthesis, double-sided spectrum.

```
sr = 44100.
f1 = 3000
f2 = 200.
N = int((sr/2 - f1)/f2)

t = pl.arange(0,sr)
w = 2*pl.pi*f1*t/sr
o = 2*pl.pi*f2*t/sr
a = 0.4
if a > 10**(-3./N):
    a = 10**(-3./N)

sinw = pl.sin(w)
den = 1.- 2*a*pl.cos(o) + a*a
scal = pl.sqrt((1. - a*a)/(1+a*a))
s = scal*(1-a*a)*sinw/den
```

The Faust implementation of these algorithms follows closely the Python examples. As an example, we give a full program based on the non-band-limited double-sided formula of Eq. 4.11 in Listing 4.6. Likewise, many other methods in this section can be implemented in Faust by applying directly the recipes provided.

Listing 4.6 Faust instrument based on Eq. 4.11.

```
vol = hslider("amp", 0.1, 0, 1, 0.001);
freq1 = hslider("freq1 [unit:Hz]",440,110,1760,1);
freq2 = hslider("freq2 [unit:Hz]",440,110,1760,1);
a = hslider("a", 1,0,1,0.001);

pi = 3.141592653589793;
sr = 44100;
```

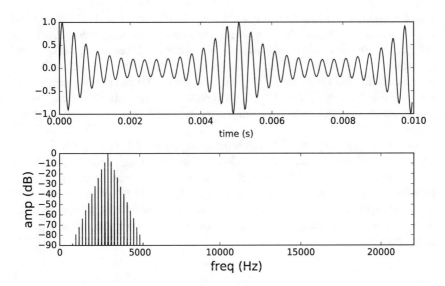

Fig. 4.5 Plot of a waveform and its spectrum as generated by the program in listing 4.5.

```
mod1(a) = a - floor(a);
incr(freq) = freq / float(sr);
phasor(freq) = incr(freq) : (+ : mod1) ~ _ ;
w = 2*pi*phasor(freq1);
th = 2*pi*phasor(freq2);
sig = \sin(w)*(1. - a*a)/(1. - 2*a*cos(th) + a*a);
process = vol*sig;
```

4.3 Frequency and Phase Modulation

Frequency Modulation (FM) was pioneered by John Chowning in the early 1970s
[15]. It is a special case of summation formulae that has a very straightforward implementation with two oscillators. A version of this algorithm, called Phase Modulation (PM), turns out to be more malleable to use. For this reason, we will concentrate
on this particular version of the technique in the discussion that follows.

The expression for the simple form of PM, with one carrier, and one modulator,
is given by the expression:

$$\cos(\omega_c + I\sin(\omega_m)) = \sum_{n=-\infty}^{\infty} J_n(I)\cos(\omega_c + n\omega_m) \qquad (4.13)$$

where $J_n(I)$ are Bessel functions of the first kind (of order n), and $\omega_c = 2\pi f_c t$ and $\omega_m = 2\pi f_m t$ are known as the carrier and modulator frequencies, respectively. As we can see, the principles involved here are not too different from those of the generalised summation formulae discussed in the previous section. With PM, we have a double-sided spectrum that might span both negative and positive frequencies depending on the values of f_c and f_m. An important difference is the presence of the Bessel-function scaling, which determines the amplitude of each component in the spectrum

These functions depend on I, which is the amplitude of the modulator sinusoid. This is also called the *index* of modulation. For low I, the high-order $J_n(I)$ are zero or close to zero. As I rises, these Bessel functions tend to increase, and then fluctuate between positive and negative values (see Fig. 4.6). Unlike the cases we have seen before, the spectral rolloff does not increase/decrease linearly with the change in the index of modulation. Instead there is a considerable amount of fluctuation of partial amplitudes as I is varied.

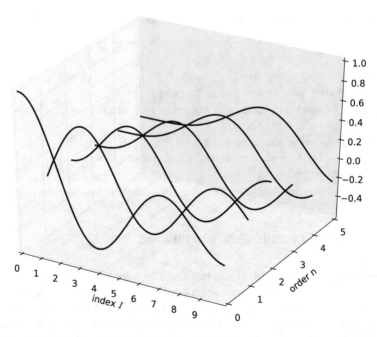

Fig. 4.6 Bessel functions of orders 0 to 5. Note how they oscillate between positive and negative values as the index of modulation increases.

From Eq. 4.13, we can see that PM features spectra that are made up of the sums and differences of the carrier and modulator frequencies (plus the carrier itself). Each one of these $f_c \pm f_m$ frequencies is called a *sideband*, lying as it does on one side or the other of the carrier. Sidebands are scaled by Bessel functions, with negative-order functions obeying the following relationship:

$$J_{-n}(I) = (-1)^n J_n(I) \tag{4.14}$$

Cosine components on the negative side of the spectrum are folded back on the positive side, since $\cos(-x) = \cos(x)$. If the components were sine waves, then they would have reversed sign $(\sin(-x) = -\sin(x))$. This is the case for a variation of PM where the carrier is in the sine phase[3]. The $f_c : f_m$ ratio will determine whether the spectrum is harmonic or inharmonic. As we have seen in the other cases of summation formulae, if involves small whole numbers, we have harmonic partials, otherwise the spectral components will be perceived as inharmonic. The amplitude of each component in the spectrum can be worked out from Eq. 4.13 above and the values of $J_n(I)$ for a given k. Most of the spectral energy will be concentrated in the first $I + 1$ sidebands, although the waveform is technically non band limited.

The PM algorithm is implemented in Listing 4.7. Time-varying spectra are possible (as in all of the other techniques studied here). In this case, all we need to do is to dynamically shape the modulation index (see also Listings B.6 and B.7).

Listing 4.7 PM synthesis

```
import pylab as pl

sr = 44100.
fc = 500.
fm = 1000.

t = pl.arange(0,sr)
w = 2*pl.pi*fc*t/sr
o = 2*pl.pi*fm*t/sr
I = 3.5
s = pl.cos(w + I*pl.sin(o))
```

A number of other modulation combinations are possible including complex (where we use a modulator with more than one harmonic) [13, 92], feedback [102] (where the output is fed back to modulate the frequency; see Chap. 5), and multi-carrier modulation (where several carriers are modulated by the same source).

4.4 Asymmetrical PM synthesis

Asymmetrical PM [79] is a variation on the original algorithm. Here, we ring modulate the original PM signal by an exponential. A new parameter is introduced, controlling spectral symmetry, which allows the peaks to be dislocated above or below

---

[3] This is the case studied in Chowning's original paper.

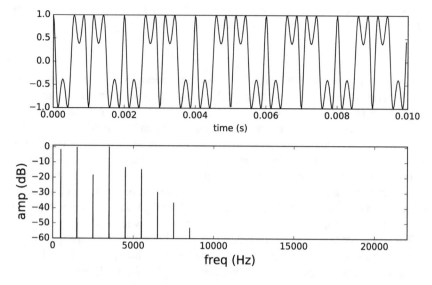

Fig. 4.7 Plot of a phase modulation waveform and its spectrum as generated by the program in listing 4.7.

the carrier frequency. The expression for this technique (excluding a normalisation factor) is

$$
\exp\left(\frac{I}{2}\left(r-\frac{1}{r}\right)\cos\left(\omega_m\right)\right) \times \sin\left(\omega_c+\frac{I}{2}\left(r+\frac{1}{r}\right)\sin\left(\omega_m\right)\right)
$$
$$
= \sum_{n=-\infty}^{\infty} r^n J_n\left(I\right)\sin\left(\omega_c+n\omega_m\right)
\tag{4.15}
$$

where $J_n(I)$ are Bessel functions of the first kind; and $\omega_c = 2\pi f_c t$ and $\omega_m = 2\pi f_m t$ are the carrier and modulator frequencies, respectively[4]. The new symmetry parameter is r, with $r < 1$ moving the spectral peak below the carrier frequency ω_c and $r > 1$ placing it above. This is an enhancement of PM with a few extra multiplies and a couple of extra function calls.

The exponential expression needs to be scaled, which can be achieved by dividing it by $\exp(\frac{I}{2}[r - \frac{1}{r}])$. From this and Eq. 4.15, we can program the technique as shown in Listing 4.8 (see also Listing B.8).

Listing 4.8 Asymmetric FM.

```
import pylab as pl
```

---

[4] For convenience, we notate e^x as $\exp(x)$ here and in other formulae as needed.

```
sr = 44100.
fc = 500.
fm = 1000.

t = pl.arange(0,sr)
w = 2*pl.pi*fc*t/sr
o = 2*pl.pi*fm*t/sr
I = 3.5
r = 2.

k1 = 0.5*I*(r - 1./r)
k2 = 0.5*I*(r + 1./r)
scal = 1./pl.exp(0.5*I*(r - 1./r))
s = scal*pl.exp(k1*pl.cos(o))*pl.sin(w + k2*pl.sin(o))
```

A plot of the waveform and spectrum for asymmetric PM is shown in Fig. 4.8. Comparing this with the example in Fig. 4.7, it is possible to see how the energy has shifted upwards in the spectrum as a result of the use of a parameter r above unity.

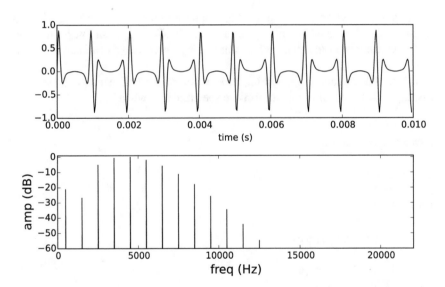

Fig. 4.8 Plot of an asymmetric phase modulation waveform and its spectrum as generated by the program in listing 4.8.

4.5 Phase-Aligned Formant Synthesis

Another method based on a summation formula[5], producing a double-sided spectrum, is provided by the Phased-Aligned Formant (PAF) algorithm [84]. The main interest here is creating formant regions around a centre frequency. The basic expression for PAF is given by

$$\frac{1+g}{1-g} \times f\left(\frac{2\sqrt{g}}{1-g}\sin\left(\frac{\omega_m}{2}\right)\right)\cos(\omega_c) = \sum_{n=-\infty}^{\infty} g^{|n|}\cos(\omega_c + n\omega_m) \qquad (4.16)$$

with

$$f(x) = \frac{1}{1+x^2} \qquad (4.17)$$

and

$$g = \exp(\frac{f_c}{B}) \qquad (4.18)$$

A transfer function $f(x)$ is applied to a sinusoid tuned to $\omega_m = 2\pi f_m t$. This is ring-modulated by another sinusoid at $\omega_c = 2\pi f_c t$. When f_c is an integer multiple of f_c, the result is a signal whose fundamental is f_m containing a formant centred at f_c.

Listing 4.9 contains a Python implementation of the PAF expression (see also Listing B.9). The resulting waveform and spectrum are shown in Fig. 4.9.

Listing 4.9 PAF.

```
import pylab as pl

sr = 44100.
fc = 3000.
fm = 100.

t = pl.arange(0,sr)
w = 2*pl.pi*fc*t/sr
o = 2*pl.pi*fm*t/sr
bw = 4000.

def f(x): return 1./(1.+x*x)
g = pl.exp(fc/bw)
k1 = (1.+g)/(1.-g)
k2 = 2*pl.sqrt(g)/(1.-g)
s = k1*f(k2*pl.sin(o))*pl.cos(w)
```

---

[5] In fact, this is very similar to the generalised double-sided (non-band-limited) formula given in Eq. 4.11.

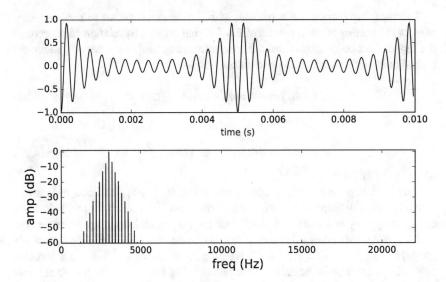

Fig. 4.9 Plot of a PAF waveform and its spectrum as generated by the program in listing 4.9.

This formula is particularly significant as it can be implemented very efficiently using wave-shaping methods, which are discussed in Sect. 4.8. In this case, instead of calculating $f(x)$ directly, we can implement the operation using table lookup.

4.6 Split-Sideband Synthesis

Split-sideband synthesis (SpSB) [58] is a closed-form summation method which uses PM as a starting point. It produces four independent outputs, containing the resulting complex spectra in separate sideband groups: lower-odd, lower-even, upper-odd and upper-even. SpSB synthesis is based on two steps: (i) production of independent odd and even sidebands; and (ii) separation of the lower and upper sideband groups.

The method for producing the even and odd sideband components uses the following closed forms:

$$\cos(I\sin(\omega)) = J_0(I) + 2\sum_{n=1}^{\infty} J_{2n}(I)\cos(2n\omega) \qquad (4.19)$$

and

$$\sin(I\sin(\omega)) = 2\sum_{n=1}^{\infty} J_{2n-1}(I)\sin(2[n-1]\omega) \qquad (4.20)$$

The resulting spectra are harmonic, with the fundamental set to the frequency $\omega = 2\pi ft$ (missing in the spectrum of Eq. 4.19 but implied by relationships beetwen the higher harmonics). To centre the spectrum at any frequency, we ring modulate both signals with a sinusoidal wave whose frequency is f_c:

$$s_{even}(t) = \sin(\omega_c t) \times \cos\left(I\sin(\omega_m t)\right) \tag{4.21}$$

and

$$s_{odd}(t) = \sin(\omega_c t) \times \sin\left(I\sin(\omega_m t)\right) \tag{4.22}$$

where $\omega_c = 2\pi f_c$ and $\omega_m = 2\pi f_m$.

This method generates two sideband sets: even, $\omega_c \pm 2n\omega_m$, and odd, $\omega_c \pm 2(n-1)\omega_m$. Separating these two groups into further two, placed above and below the carrier frequency is a matter of modifying the ring modulation expression in eqs. 4.21 and 4.22. Instead of using a real-valued carrier with a doubled-sided spectrum (negative and positive frequencies[6]), we employ an *analytic* signal. This contains either the positive or the negative spectrum only. The implication is that we are now be dealing with complex-valued signals.

In this case the sine wave carrier $c(t)$ is defined as

$$c(t) = \sin(\omega_c t) - j\cos(\omega_c t) \tag{4.23}$$

To implement the complex-signal ring-modulation, we will need to employ an all-pass filter that shifts the phase of a signal by 1/4 of a cycle ($\pi/2$), putting it into a *quadrature* relationship with the original signal. This is achieved by an operation called the *Hilbert* transform H, which can be used to produce complex-valued signals with a single-sided spectrum:

$$z_{positive}(t) = s(t) + jH\left(s(t)\right) \tag{4.24}$$

$$z_{negative}(t) = s(t) - jH\left(s(t)\right) \tag{4.25}$$

With single-sideband modulation instead of the ring modulation in eqs. 4.21 and 4.22, it is possible to implement SpSB. The upper and lower sidebands are produced by the following matrix expression:

$$\begin{bmatrix} s_{upper,even}(t) & s_{upper,odd}(t) \\ s_{lower,even}(t) & s_{lower,odd}(t) \end{bmatrix} = \begin{bmatrix} \sin(\omega_c t) & \cos(\omega_c t) \\ \sin(\omega_c t) & -\cos(\omega_c t) \end{bmatrix} \times \\ \begin{bmatrix} \cos\left(I\sin(\omega_m t)\right) & \sin\left(I\sin(\omega_m t)\right) \\ H\left(\cos\left(I\sin(\omega_m t)\right)\right) & H\left(\sin\left(I\sin(\omega_m t)\right)\right) \end{bmatrix} \tag{4.26}$$

---

[6] Real-valued signals, such as audio waveforms, have a spectrum composed of both positive and negative frequencies, symmetric about 0 Hz, see Chap. 7 for further details.

This is implemented in Listing 4.10, where the matrix multiplication of Eq. 4.26 is written out as four separate lines. The quadrature filter is provided by the ScipPy function `hilbert`, which produces a complex number whose imaginary part is what we are interested in. A plot of the four SpSB spectra is shown in Fig. 4.10.

Listing 4.10 SpSB synthesis.

```python
import pylab as pl
from scipy.signal import hilbert


sr = 44100.
fc = 3000
fm = 500.

t = pl.arange(0,sr)
w = 2*pl.pi*fc*t/sr
o = 2*pl.pi*fm*t/sr
I = 3.5

sod = pl.sin(I*pl.sin(o))
sev = pl.cos(I*pl.sin(o))
qsod = pl.imag(hilbert(sod))
qsev = pl.imag(hilbert(sev))
sinw = pl.sin(w)
cosw = pl.cos(w)

upper_even = sev*sinw + qsev*cosw
upper_odd  = sod*sinw + qsod*cosw
lower_even = sev*sinw - qsev*cosw
lower_odd  = sod*sinw - qsod*cosw
```

4.7 Modified FM Synthesis

Modified FM synthesis (ModFM) is another variation on the original PM algorithm [53]:

$$\exp(k\cos(\omega_m) - k)\cos(\omega_c) =$$
$$\frac{1}{e^k}\left(I_0(k)\cos(\omega_c) + \sum_{n=1}^{\infty} I_n(k)\left[\cos(\omega_c + n\omega_m) - \cos(\omega_c - n\omega_m)\right]\right) \quad (4.27)$$

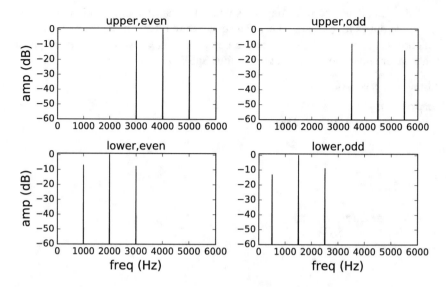

Fig. 4.10 The four SpSB spectra as generated by the program in listing 4.10.

where $\omega_c = 2\pi f_c t$ and $\omega_m = 2\pi f_m t$ are the carrier and modulator frequencies; and $I_n(k)$ are *modified* Bessel functions of the first kind[7]. This is the basic difference between FM/PM and ModFM. The modified Bessel functions have two advantages over the ordinary Bessel functions for this particular application:

1. They are unipolar.
2. $I_n(k) > I_{n+1}(k)$, which means that dynamic spectrum shaping is much more natural here.

In particular, the scaled modified Bessel functions do not exhibit the problematic fluctuations observed in the behaviour of Bessel functions. That very unnatural-sounding characteristic of FM disappears in ModFM. A plot of modified Bessel functions of orders 0 to 5 is shown in Fig. 4.11.

The straight implementation of this formula is shown in Listing 4.11. As indicated by Eq. 4.27, a scaling factor $\frac{1}{e^k}$ is used to normalise the waveform (see also Listing B.10).

Listing 4.11 ModFM synthesis.

```
import pylab as pl

sr = 44100.
fc = 500.
```

---

[7] To avoid confusion, we will denote the index of modulation by k here, rather than I as originally used.

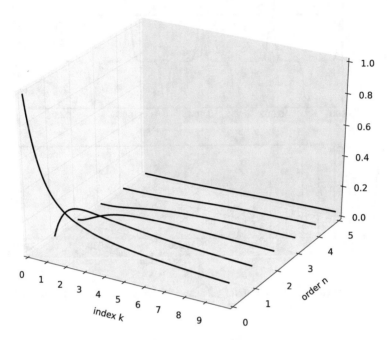

Fig. 4.11 Modified Bessel functions of orders 0 to 5. Unlike the original Bessel functions, these do not oscillate and are unipolar.

```
fm = 1000.

t = pl.arange(0,sr)
w = 2*pl.pi*fc*t/sr
o = 2*pl.pi*fm*t/sr
I = 3.5
scal = 1./pl.exp(I)
s = scal*pl.exp(I*pl.cos(o))*pl.cos(w)
```

Figure 4.12 displays a ModFM spectrum and waveform.

With ModFM, as with other summation formula methods, it is possible to realise typical low-pass filter effects by varying the index of modulation k. The algorithm also allows us to emulate a combination of a source with a band-pass filter [51, 54]. In this scenario, we select f_m as the centre of resonance and f_c as the fundamental frequency, and we can calculate the index of modulation k from a bandwidth parameter (Listings 4.12 and B.11). This is a very efficient alternative to PAF and other formant techniques, such as, for instance, FOF [90] synthesis (Fig. 4.13).

Listing 4.12 ModFM formant synthesis.

```
import pylab as pl
```

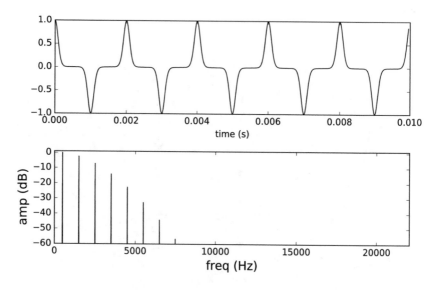

Fig. 4.12 Plot of a ModFM waveform and its spectrum as generated by the program in listing 4.11.

```
sr = 44100.
fc = 3000.
fm = 100.

t = pl.arange(0,sr)
w = 2*pl.pi*fc*t/sr
o = 2*pl.pi*fm*t/sr

bw = 1000
g = pl.exp(-fm/(.29*bw))
g2 = 2*pl.sqrt(g)/(1.-g)
I = g2*g2/2
scal = 1./pl.exp(I)
s = scal*pl.exp(I*pl.sin(o))*pl.cos(w)
```

4.7.1 Extended ModFM

The original ModFM expression can be extended to modify the symmetry of its output spectrum, somewhat as in Sect. 4.4. If we employ an extra phase modulation

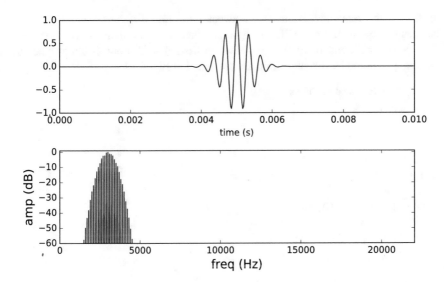

Fig. 4.13 The plot of a resonant ModFM waveform and its spectrum as generated by the program in listing 4.12.

component, plus two parameters, r and s (with ranges $0 \leq r \leq 1$ and $-1 \leq s \leq 1$), we have the following expression (which excludes the normalisation factor e^{-rk}):

$$
\begin{aligned}
\exp(rk\cos(\omega_m))\cos(\omega_c + sk\sin(\omega_m)) = \\
\sum_{a=-\infty}^{\infty} I_a(rk)\cos(a\omega_m) \sum_{b=-\infty}^{\infty} J_b(sk)\cos(\omega_c + b\omega_m) = \\
\frac{1}{2} \sum_{a=-\infty}^{\infty}\sum_{b=-\infty}^{\infty} I_a(rk)J_b(sk)\left[\cos(\omega_c + (a+b)\omega_m) + \cos(\omega_c - (a-b)\omega_m)\right]
\end{aligned}
\tag{4.28}
$$

This is the closed form of a complex summation, involving both ordinary and modified Bessel functions. However, it has a simpler expansion in the extreme ranges of its r and s parameters. With $s = 0$ and $r = 1$, we have the original ModFM expression. If $r = 0$ and $s \neq 0$, we have the PM formula. If $r = s = 1$, the expansion shows a single-sided spectrum with no Bessel functions as scaling factors:

$$
exp(k\cos(\omega_m))\cos(\omega_c + k\sin(\omega_m)) = \sum_{n=0}^{\infty} \frac{k^n}{n!} \cos(\omega_c + n\omega_m)
\tag{4.29}
$$

In this case the spectral envelope peaks at sideband $n = k$. Its shape is close to that of the original ModFM ($s = 0$ and $r = 1$), as the scaling factors in Eq. 4.29 approximate modified Bessel functions that are shifted in frequency. The other extreme case, $s = -r = -1$, also generates single-sided spectrum, now on the lower side of the carrier frequency. Transitions between these four different spectral envelope shapes can be made by modifying these parameters within their allocated ranges. The code for Extended ModFM is shown in Listing 4.13. Four different types of spectra generated by this algorithm are presented in Fig. 4.14.

Listing 4.13 Extended ModFM synthesis.

```
import pylab as pl

sr = 44100.
ffc = 5000
fm = 500.

t = pl.arange(0,sr)
w = 2*pl.pi*fc*t/sr
o = 2*pl.pi*fm*t/sr
I = 3.5
s = 1
r = 1
scal = 1./(pl.exp(I))
sig = scal*pl.exp(I*r*pl.cos(o))*pl.cos(w + I*s*pl.sin(o))
```

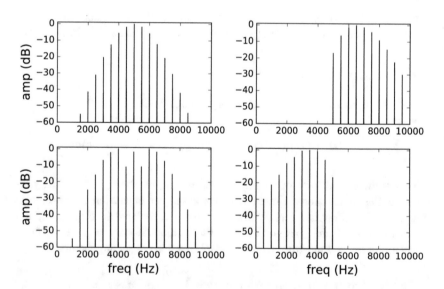

Fig. 4.14 Spectra generated by the Extended ModFM algorithm. Top left shows the result of setting $r = 1$ b and $s = 0$ (original ModFM); bottom left uses $s = 1$ and $r = 0$ (PM); top right, $s = r = 1$ (upper sidebands); and bottom right, $s = -r = -1$ (lower sidebands).

4.8 Wave-Shaping Synthesis

Wave-shaping synthesis is based on non-linear distortion of the amplitude of a signal [2, 61]. This is achieved by mapping an input, generally a simple sinusoidal one, using a function that will modify it into a desired output waveform. The amount of distortion, and the resulting spectrum, can be controlled by changing the amplitude of the input signal.

General-purpose wave shaping is just the application of a function to an input signal, possibly distorting it if the function is non-linear (see Fig. 4.15). It takes an input sinusoid, maps it using function lookup and then applies the required gain to normalise it, as implemented in Listing 4.14 (see also Listing B.12).

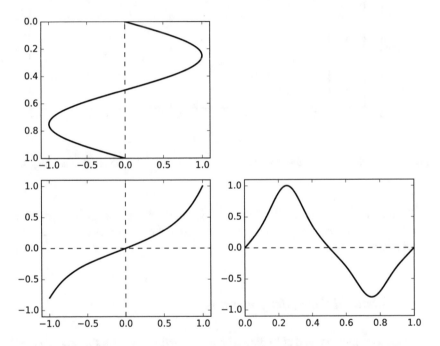

Fig. 4.15 The principle of wave shaping. The plot at the top shows the input to the wave shaper, whose transfer function is shown at the bottom left, and the resulting output on the right.

Listing 4.14 Wave-shaping synthesis.

```
import pylab as pl

sr = 44100.
fc = 500.
```

```
t = pl.arange(0,sr)
w = 2*pl.pi*fc*t/sr

def f(x): return x*0.7 + 0.1*x**2 + 0.2*x**3 +
          0.05*x**4 +  0.4*x**5
scal = 1./abs(f(1.0))
s = scal*f(pl.sin(w))
```

The spectrum of the waveform produced by this program is shown in Fig. 4.16. The transfer function used here is the same as in the earlier example (Fig. 4.15).

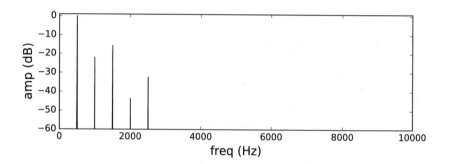

Fig. 4.16 Plot the spectrum of a waveform generated by the wave shaping program in listing 4.14.

By changing the amplitude of the driving sinusoid, we can generate dynamic spectra. It will be necessary however to compensate these changes in volume with a suitable scaling function.

4.8.1 Polynomial transfer functions

The most common method of finding a transfer function for wave shaping is through polynomial spectral matching. This approach is based on the principle that polynomial functions will produce a precisely defined band-limited spectrum when driven by a sinusoidal signal of a given amplitude. In the examples shown in figs. 4.15 and 4.16, we have used a polynomial expression to define the transfer function:

$$f(x) = 0.7x + 0.1x^2 + 0.2x^3 + 0.05x^4 + 0.4x^5 \qquad (4.30)$$

The relationship between wave shaping and closed-form summation can be demonstrated by expanding the expression $f(sin(\omega))$:

$$f(\sin(\omega)) = 0.7\sin(\omega) + 0.1\sin(\omega)^2 +$$
$$0.2\sin(\omega)^3 + 0.05\sin(\omega)^4 +$$
$$0.4\sin(\omega)^5$$
$$= 0.7\sin(\omega) + 0.5 - 0.05\cos(2\omega) +$$
$$0.15\sin(\omega) - 0.05\sin(3\omega) +$$
$$0.019 - 0.025\cos(2\omega) + 0.006\cos(4\omega) + \qquad (4.31)$$
$$0.25\sin(\omega) - 0.125\sin(3\omega) + 0.025\sin(5\omega)$$
$$= 0.244 + 1.1\sin(\omega) - 0.075\cos(2\omega) +$$
$$0.174\sin(3\omega) + 0.006\cos(4\omega) + 0.025\sin(5\omega)$$
$$= \sum_{n=0}^{5} A_n\cos(n\omega + \phi_n)$$

Eq. 4.31 shows how the wave-shaping expression is a closed form of a certain band-limited spectrum. Other things can be inferred from this. First, even-order coefficients (connected to powers of even numbers) contribute only to even harmonics, and odd-order to odd. Second, the highest component of the spectrum will never exceed the order of the polynomial, so we can guarantee a band-limited output.

Specifying a polynomial directly may not always be very practical. As observed in Eq. 4.31, the expansion of the expression can become very complicated to work out. However, if we have a given steady-state spectrum that we would like to generate, it is possible to use the following relation [61]:

$$T_n(\cos(\omega)) = \cos(n\omega) \qquad (4.32)$$

where $T_n(x)$ is a Chebyshev polynomial of the first kind. These can be added together to construct a polynomial function that will produce a given spectrum if driven by a cosine wave. So, if we start with a series of harmonic amplitudes a_n, we can find the polynomial $f(x)$

$$f(x) = \sum_{n=0}^{N} a_n T_n(x) \qquad (4.33)$$

and generate a signal with

$$f(\cos(\omega)) = \sum_{n=0}^{N} a_n\cos(n\omega) \qquad (4.34)$$

The program in Listing 4.15 demonstrates this approach. We use the `chebyval` function from NumPy to implement Eq. 4.33 and create a polynomial function `f()` that we apply to a cosine input. The program matches the amplitude spectrum of a sawtooth wave (Fig. 4.17). However since it is made of a sum of cosines rather than sine waves, the actual waveform shape is different.

Listing 4.15 Spectral matching with Chebyshev polynomials.

```
import pylab as pl
import numpy.polynomial.chebyshev as chb

def chebygen(a):
  return lambda x: chb.chebval(x,a)

sr = 44100.
fc = 500.
N = int(sr/(2*fc))
amps = pl.zeros(N)
for i in range(1,N):
  amps[i] = 1./i

t = pl.arange(0,sr)
w = 2*pl.pi*fc*t/sr

f = chebygen(amps)
scal = 1./abs(f(1.0))
s = scal*f(pl.cos(w))
```

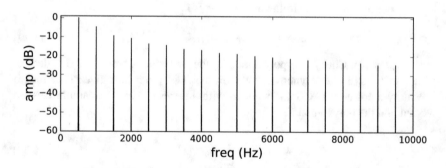

Fig. 4.17 Plot the spectrum of a waveform generated by the wave-shaping program in listing 4.15.

When the cosine amplitude is below unity, the resulting spectrum will not match the desired harmonic amplitudes. Similarly to PM, the spectral evolution is not linear as we increase the amount of distortion from 0 to 1.

This method produces a cosine-phase spectrum. Similarly, we are able to generate a signal with harmonics in the sine phase. For this, the following expression applies [61]:

$$\sin(\omega)U_{n-1}(\cos(\omega)) = \sin(n\omega) \tag{4.35}$$

where $U_n(x)$ is a Chebyshev polynomial of the second kind. So, again, we can start with sine wave amplitudes b_n and build a wave shaper function $g(x)$,

$$g(x) = \sum_{n=1}^{N} b_n U_{n-1}(x) \tag{4.36}$$

with which we generate the signal

$$\sin(\omega)g(\cos(\omega)) = \sum_{n=1}^{N} b_n \sin(n\omega) \tag{4.37}$$

Using Eq. 4.35, the code in Listing 4.16 demonstrates how to generate a signal whose amplitudes and phases match a sawtooth wave (see Fig. 4.18). The numpy.polynomial.chebyshev package does not support the generation of Chebyshev polynomials of the second kind. However, we can create the wave shaper by hand in a NumPy array and then supply a function based on it to the program.

Listing 4.16 Generating sine-phase harmonics.

```
import pylab as pl
import scipy.special as sp

def chebygen(amps):
    N = 40000
    p = pl.arange(-N,N+1)
    c = pl.zeros(len(p))
    n = 0
    for i in amps:
        c += sp.eval_chebyu(n-1,p/N)*i
        n += 1
    return pl.vectorize(lambda x: c[int(x*N+N)])

sr = 44100.
fc = 100.
N = int(sr/(2*fc))
amps = pl.zeros(N)
for i in range(1,N):
    amps[i] = 1./i

t = pl.arange(0,sr)
w = 2*pl.pi*fc*t/sr

g = chebygen(amps)
scal = 2/pl.pi
s = scal*pl.sin(w)*g(pl.cos(w))
```

This allows us to combine two wave shapers to synthesise a steady-state spectrum containing n harmonics with amplitudes A_n and phases ϕ_n. For this, two wave-

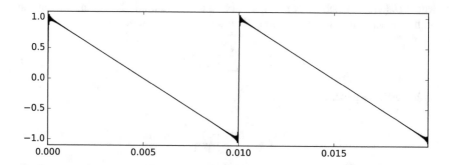

Fig. 4.18 Plot the sawtooth waveform generated by the wave-shaping program in listing 4.16.

shaping functions $f(x)$ and $g(x)$ need to be constructed as per eqs. 4.33 and 4.36, respectively. The amplitudes a_n and b_n are then defined as

$$a_n = A_n \cos(\phi_n) \tag{4.38}$$

$$b_n = -A_n \sin(\phi_n) \tag{4.39}$$

The outputs of the two wave shapers can then be combined to generate the waveform:

$$f(\cos(\omega)) + \sin(\omega)g(\cos(\omega)) = \sum_{n=0}^{N} A_n \cos(n\omega + \phi_n) \tag{4.40}$$

This provides a very compact closed form for the Fourier series of a steady-state waveform. However, as noted above, generating dynamic spectra with polynomial wave shapers can be problematic as the changes in the individual harmonic amplitudes are not linear with changes in the amplitude of the driving sinusoid. This can often produce non-smooth spectral evolutions, similar to the 'wobble' seen in PM synthesis.

4.8.2 Hyperbolic tangent wave shaping

For certain wave shapes, it is possible to use a variety of other functions, from the trigonometric, hyperbolic, etc. repertoire. For instance, we have seen how the cosine and sine functions can be used as wave shapers in SpSB synthesis. Some of these functions can provide smoother spectral transitions than the polynomial case. As an example, we can look at the use of hyperbolic tangent transfer functions to generate nearly band-limited square and sawtooth waves, which could be employed as sources for filtering.

A non-band-limited square wave can be generated through the use of the *signum*() function, mapping a varying bipolar input. This piecewise function outputs 1 for all non-zero positive input values, 0 for a null input and -1 for all negative arguments. As it clips the signal, it generates lots of aliased components. The main cause of these is the discontinuity at 0, where the output moves from fully negative to fully positive. which needs to be smoothed.

The hyperbolic tangent [52] has a smooth transition at that point, but also preserves some of the necessary clipping aspect. Driving this function with a sinusoidal input produces a quasi-band-limited signal,

$$sq(\omega) = \tanh(k\sin(\omega)) \tag{4.41}$$

where $\omega = 2\pi ft$ is the fundamental frequency. In order to keep aliasing at bay, the index of distortion k can be estimated roughly as $k = 10^4/(f_0\log_{10}(f_0))$. As with all types of wave shaping, the amplitude of the input signal determines the signal bandwidth proportionally. To keep a steady output amplitude, while varying the spectrum, we apply amplitude-dependent scaling as before. The hyperbolic wave shaping program is shown in Listing 4.17, and its generated signal is plotted in Fig. 4.19.

Listing 4.17 Hyperbolic wave shaping.

```
import pylab as pl

sr = 44100.
fc = 500.

t = pl.arange(0,sr)
w = 2*pl.pi*fc*t/sr

k = 10000./(fc*pl.log10(fc))
scal = 1./pl.tanh(k)
s = scal*pl.tanh(k*pl.sin(w))

pl.figure(figsize=(8,5))
```

To generate a sawtooth wave instead, we can take our square signal and apply the following expression:

$$\frac{1}{2}sq(\omega)(\cos(\omega)+1) \tag{4.42}$$

By heterodyning this with a cosine wave, it is possible to generate the missing even components that make up the sawtooth signal. The amplitude of the second harmonic will be off by about 2.5 dB, but the higher harmonics will be close to what we expect [52]. This is demonstrated in Listing 4.18 and Fig. 4.20 (see also Listing B.13).

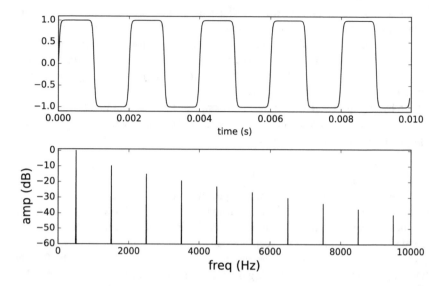

Fig. 4.19 A square wave and its spectrum, as generated by the wave-shaping program in listing 4.17.

Listing 4.18 Sawtooth wave generation based on wave shaping.

```
import pylab as pl

sr = 44100.
fc = 500.

t = pl.arange(0,sr)
w = 2*pl.pi*fc*t/sr

k = 10000./(fc*pl.log10(fc))
scal = 1./pl.tanh(k)
sq = scal*pl.tanh(k*pl.sin(w))
saw = 0.5*sq*(pl.cos(w) + 1.)
```

4.9 Phase Distortion Synthesis

Phase Distortion (PD) Synthesis [47] is effectively complex-modulation[8] PM formulated in a different way. In this method we apply a non-linear transfer function

---

[8] Complex modulation in this context refers to the use of a non-sinusoidal modulating source.

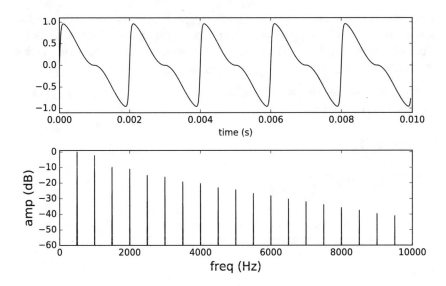

Fig. 4.20 A sawtooth wave and its spectrum, as generated by the wave-shaping program in listing 4.18.

to the phase increment of a sinusoidal oscillator. The basic form of PD synthesis is defined by eqs. 4.43 and 4.44:

$$s(t) = -\cos(2\pi\phi_{pd}(\phi(t)))$$

(4.43)

$$\phi_{pd}(x) = \begin{cases} \frac{1}{2}\frac{x}{d}, & x < d \\ \frac{1}{2}[1 + \frac{(x-d)}{(1-d)}], & x \geq d \end{cases}$$

(4.44)

where t is time in samples. In this case, Eq. 4.44 is a phase shaper acting on an input signal, $\phi(t)$. This is a geometric sawtooth wave with frequency f_0 and sampling rate f_s in Hz

$$\phi(t) = \left[\frac{f_0}{f_s} + \phi(t-1)\right] \bmod 1$$

(4.45)

which is the phase signal used in a standard table lookup oscillator, for instance (see also Listing 3.6 and Fig. 3.3 in Chap. 3).

In this phase shaper, the point of inflection d in Eq. 4.44 determines the brightness, the number of harmonics and the waveform. If d is close to 0 or to 1, the signal will be very bright and more likely to introduce audible aliasing. At $d = 0.5$, there is no change in the phase signal as the shaper function is linear. By varying d, we can get a low-pass filter effect. It is possible. of course, to add other inflection points, which would produce different wave shapes (approximating square and pulse waves for instance). The following program demonstrates the algorithm, where we define a phase-shaping function to take in the point of inflection d (Listing 4.19, which is used to produce Fig. 4.21):

Listing 4.19 Phase Distortion synthesis program.

```
import pylab as pl

sr = 44100.
fc = 500.
t = pl.arange(0,sr)

def pd(x,d):
        if x < d: return (0.5*x)/d
        else:     return 0.5*(x-d)/(1-d) + 0.5

def phasor(ph,freq,sr):
    incr = freq/sr
    ph += incr
    return ph - pl.floor(ph)

s = pl.zeros(sr)
ph = 0.0
for n in t:
    ph = phasor(ph,fc,sr)
    s[n] = pl.cos(2*pl.pi*pd(ph,0.9))
```

4.9.1 Vector phase shaping

In PD, the inflection point d controls the horizontal position of the bending point of the phase-shaping function. In this case, the vertical position of the bend is fixed at 0.5. Vector Phase-Shaping (VPS) [37] releases this constraint, expressing the inflection point as a two-dimensional vector

$$p = (d, v), \tag{4.46}$$

where d, $0 \leq d \leq 1$, is the horizontal, and v is the vertical position of the bending point. The two-dimensional phase distortion function is given as

$$\phi_{vps}(x) = \begin{cases} \frac{vx}{d}, & x < d \\ (1-v)\frac{(x-d)}{(1-d)} + v, & x \geq d \end{cases} \tag{4.47}$$

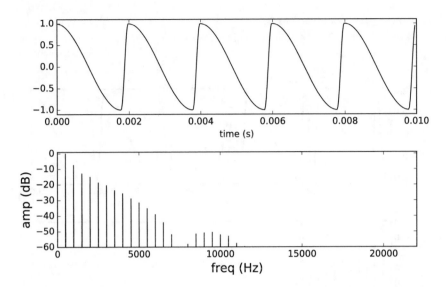

Fig. 4.21 A sawtooth-like wave and its spectrum, as generated by the phase distortion program in listing 4.19.

VPS reduces to the original PD form 4.44 when $v = 0.5$ but, in general, it is capable of producing a wider variety of spectra. The example in Listing 4.20 uses a $(0.3, 1.5)$ vector, producing the signal in Fig. 4.22.

Listing 4.20 VPS synthesis program.

```
import pylab as pl

sr = 44100.
fc = 500.
t = pl.arange(0,sr)

def vps(x,d,v):
        if x < d: return (v*x)/d
        else:     return (1-v)*(x-d)/(1-d) + v

def phasor(ph,freq,sr):
    incr = freq/sr
    ph += incr
    return ph - pl.floor(ph)

s = pl.zeros(sr)
ph = 0.0
for n in t:
```

```
ph = phasor(ph,fc,sr)
s[n] = pl.cos(2*pl.pi*vps(ph,0.3,1.5))
```

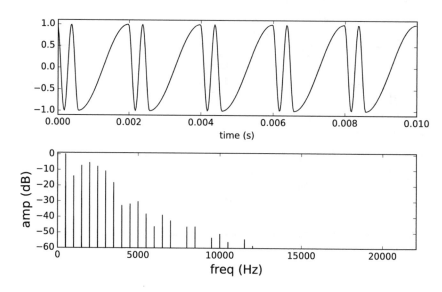

Fig. 4.22 A VPS wave and its spectrum, as generated by the program in listing 4.20.

4.10 Instrument Design

Closed-form summation formulae synthesis allows a number of interesting ways to design instruments. They pack the means to generate various types of spectra dynamically into compact algorithms. In particular, by manipulating only a few parameters, it is possible to completely transform the output signal. In the examples that follow, we will study instrument development from three perspectives: 1) use of a single design that can produce sounds with diverse characteristics by careful parameter setting; 2) analysis of arbitrary sounds followed by synthesis through a summation formula method; and 3) modelling of source-filter methods.

4.10.1 Phase modulation

Many of the techniques explored in this chapter are very versatile and can create a variety of spectral types. As an example of this flexibility, we will look at the

case of phase modulation. It is quite remarkable that we can design a minimal PM
instrument and with this, generate various different sonorities by modifying its parameters. This is also the case with the other related methods, so the ideas contained
in this section are applicable more widely.

The instrument we will use is a Csound implementation of Eq. 4.13 and the code
in Listing 4.7. Variations on this instrument can also found in Listings B.6 and B.7.
The basic principle is to use a sine wave to modulate the phase of a cosine wave
(alternatively, we could use any other combination of sines and cosines). We will
use a single envelope that can control the amount (index) of modulation and the
amplitude. By shaping the index over time, we will be able to make the spectrum
change dynamically.

We will expose five parameters that are crucial to creating the variety of sounds
we need without introducing a multiplicity of controls that would increase the complexity of the design. It is often useful to strike a balance between flexibility and
simplicity when developing an instrument. The parameters are: the peak amplitude,
carrier frequency, maximum amount of modulation, carrier to modulator (c:m) ratio, and envelope rise (attack) time. Notice that we are not specifying the modulation
frequency directly, but doing so through the c:m ratio.

The code is shown in Listing 4.21. We use a simple trapezoid envelope (linen)
to control both the amplitude and the index of modulation. A sine wave oscillator
(oscili) is used to provide the modulation signal. The carrier phase is generated
by a phasor. The modulation is added to this signal and this is used to look up a
table (using tablei). This reads from function table -1, which is the identifier of
an internal table containing a sine wave. To make it a cosine, we give it an offset of
$\frac{1}{4}$ cycle (0.25).

The table lookup index is normalised (0 to 1) and wraps around the ends of this
range. Since the index of modulation is defined with respect to a phase in the range
from 0 to 2π, we need to scale it by $\frac{1}{2\pi}$. This is in order for us to use this parameter
in the usual ranges expected by the theory.

Listing 4.21 PM instrument with four independent parameters.

```
instr 1
 indx = p6/(2*$M_PI)
 ifc = p5
 ifm = p5/p7
 irs = p8

 anv = linen(1,irs,p3,0.1)
 afm = oscili(indx*anv,ifm)
 aph = phasor(ifc)
 acr = tablei:a(aph+afm,-1,1,0.25,1)
       out(anv*p4*acr)
endin
```

With this instrument, we will create the following four different sounds:

1. **Bass drum**: with a low carrier frequency (around 50 – 80 Hz), an irrational c:m ratio ($\sqrt{2}$), short duration and attack, and a maximm index around 10.

```
schedule(1,0,.11,0dbfs/2,50,10,sqrt(2),0.001)
```

2. **Electric bass**: fast attack, ratio close to 1 (but slightly offset to give some gentle beating), and index around 4.

```
schedule(1,0,1,0dbfs/2,cpspch(6.04),4,1.02,0.005)
```

3. **Single reed**: longer attack time, ratio close to $\frac{1}{2}$ to generate a spectrum with only odd harmonics (and also slightly offset to give some gentle beating), and index between 3 and 5.

```
schedule(1,0,1,0dbfs/2,cpspch(8.02),4,1.005/2,0.04)
```

4. **Brass**: attack time should be a little longer, ratio close to 1, and index between 3 and 5.

```
schedule(1,0,1,0dbfs/2,cpspch(8.02),6,1.001,0.08)
```

The steady-state spectra of these four sounds are shown in Fig. 4.23. An example sequence using these is presented in Listing 4.22. This is constructed in similar way to the one in the example of Listing 3.10, with a loop construct and a few random number generators. Pitch and other parameters are read from lists stored in arrays.

Listing 4.22 An example sequence using four different types of sounds created by the instrument in Listing 4.21.

```
ibp[] init 4
ibp fillarray 6.04,5.11,6.07,7.02
irat[] init 2
irat fillarray 1.002/2,1.001
icnt = 0
ichd[] init 4,4
ichd fillarray 7.07,8.02,8.06,8.11,
                7.09,8.02,8.04,8.07,
                7.07,7.11,8.02,8.06,
                7.11,8.09,7.09,8.02
istart = 1
while icnt < 360 do
  /* drum */
  schedule(1,icnt/8,0.11,0dbfs/4,rnd(50)+25,
           10+gauss(4),1.414,0.001)
  /* bass */
  schedule(1,(icnt+1)/2-rnd(0.005),0.2,
             0dbfs/5+rnd(0dbfs/20),
             cpspch(ibp[int(rnd(5))%4]),
             4,1.005,0.005)
  schedule(1,(icnt+1)/2-rnd(0.005)+0.125,0.2,
```

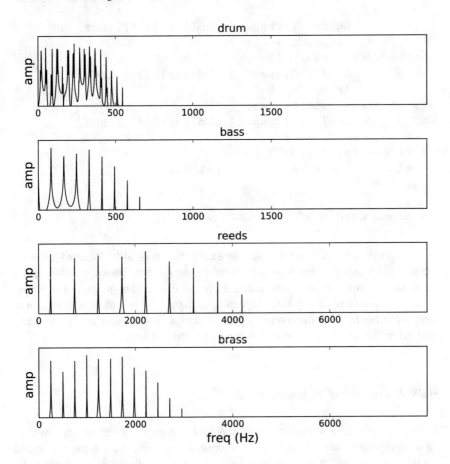

Fig. 4.23 Four different types of (steady-state) spectra generated by the instrument in listing 4.21: 'bass drum-like', 'electric bass', 'reeds-like', and 'brass-like'.

```
        0dbfs/5+rnd(0dbfs/10),
        cpspch(ibp[int(rnd(5))%4]),
        4,1.005,0.005)
/* winds synth */
idur = int(1+rnd(3))/8
irto = irat[int(rnd(3))%2]
indx = 5+gauss(3)
isel = int(rnd(5))%4
schedule(1,istart,idur,
        0dbfs/12,cpspch(ichd[isel][0]),indx,
        irto,idur*0.2)
schedule(1,istart,idur,
```

```
            0dbfs/12,cpspch(ichd[isel][1]),indx,
            irto,idur*0.2)
  schedule(1,istart,idur,
            0dbfs/12,cpspch(ichd[isel][2]),indx,
            irto,idur*0.2)
  schedule(1,istart,idur,
            0dbfs/12,cpspch(ichd[isel][3]),indx,
            irto,idur*0.2)
  irest = int(2+rnd(2))/4
  istart = round(istart+idur)+irest
  icnt += 1
od
/* close Csound after icnt/8 secs */
event_i("e",0,icnt/8)
```

It is possible to build on this simple example considerably. We could, for instance, add a separate envelope for timbre (index of modulation), possibly using more stages, exponential curves, etc. This would also lead us to expose other parameters, such as the decay time. However, as mentioned before, it is important to keep a balance between the complexity of the design, and the number of parameters, and the ability to 'play' the instrument in more intuitive way.

4.10.2 The ModFM vocoder

In this example, we will look at how closed-form summation techniques can be used in an analysis-synthesis context. In this scenario, we will extract some features of an input signal and try to reproduce these, sometimes with some modification, by driving a synthesis model with them. The choice of technique will need to match our expectations of the input signal and the aspects we are tracking off it. In this case, we will analyse the energy in the spectrum in various ranges and then try to generate a signal that reproduces these features.

In order to extract the energy in each region, we can return to some of the ideas studied in Chap. 3. There we saw that band-pass filters can be used to select a certain range of frequencies while blocking others. We saw that we can use these in parallel to filter a number of bands at the same time. While we are not interested in using the output of this filter bank directly, this can be the starting point of our analysis. What we want to extract is the energy in each band, in other words the RMS value of each filter output. The code for this becomes (with `asig` as the input):

```
asig = butterbp(asig,ifc,ibw)
asig = butterbp(asig,ifc,ibw)
krms = rms(asig)
```

This uses a 4th-order Butterworth filter and overwrites the original input, which is perfectly fine, since all we are interested is in the resulting RMS value for the

band and not the input signal itself. This takes care of one band and we can use the
same idea for all the other bands.

With the RMS data, we can control the synthesis energy in each band. Now the
question turns to the sound generation method. We want to use a technique that
can produce a series of partials around a centre frequency in a given band. From
earlier discussions, we can see that several of the summation methods can do this:
the double-sided generalised formulae, PM, PAF, and ModFM. Wave shaping with
ring modulation could also be used. For this example, we have chosen ModFM as
we can easily derive carrier and modulation frequencies from a given bandwidth,
and the spectrum is also smooth if we need to change it dynamically (although in
this particular case the actual bands will be static). The basic band design is shown
in Fig. 4.24.

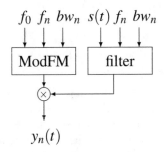

Fig. 4.24 A ModFM vocoder band n producing the output $y_n(t)$, centred at f_n ,with bandwidth
bw_n and fundamental frequency f_0, and fed by the input signal s(t).

The basic ModFM formula can be programmed as follows:

1. We derive the index of modulation from the bandwidth:

```
icor = 4.*exp(-1)
ig   = exp(-ifo/(.29*ibw*icor))
ig   = 2*sqrt(ig)/(1.-ig)
ind = ig*ig/2.
```

2. The modulation frequency is an integer multiple of the carrier frequency, closest
   to the centre frequency, $f_m = nf_0$ and $n = \lfloor \frac{f_c}{f_0} + 1 \rfloor$.
3. The whole process is driven synchronously from one phase generator, keeping
   all the sinusoids locked together, modulator sine and cosine carrier. We use table
   readers to get these signals from the default internal sine table:

```
ifn  = round(ifc/ifo)
aphs = phasor(ifo)
asin = table:a(aphs,-1,1,0,1)
aexp = exp(ind*asin - ind)
acar = tablei:a(aphs*ifn,-1,1,0.25,1)
```

```
asig = acar*aexp
```

This case works perfectly for a fixed fundamental. If we want to shape the signal over time (modulate, glide, etc.), then we need to find a way to keep the carrier locked to a multiple of the fundamental without it jumping abruptly from one frequency to another. A solution is to use two carriers at adjacent harmonics, and linearly crossfade from one to another as the centre frequency moves. The modified code looks like this:

```
kfr  = ifc/kfo
kfn  = int(kfr)
kam  = kfr - kfn
aphs = phasor(kfo)
asin = table:a(aphs,-1,1,0,1)
aexp = exp(knd*asin - knd)
acr1 = tablei:a(aphs*kfn,-1,1,0.25,1)
acr2 = tablei:a(aphs*(kfn+1),-1,1,0.25,1)
asig = (kam*acr2+(1-kam)*acr1)*aexp
```

So far, we have considered only a single band in the spectrum. We need to generate a number of them, so that we can cover a wide range of frequencies. There are two options: copy the code N times, adjusting the centre frequency and bandwidth; or programmatically spawn N bands. The first alternative is not very flexible (as we would need to fix N for all practical purposes) and would generate extremely verbose code. The second option is much handier and allows us to create a compact and very malleable instrument.

We can take advantage of the fact that instrument instances ('notes' or 'events') can be launched in parallel with schedule, make our code model a single band, and have n instances started from a loop (as in the previous example), each with a different centre frequency and bandwidth. We can go one step further and dispense with the loop altogether, by using *recursion*, where the instrument can instantiate itself.

This is possible because there is nothing to stop us from placing a schedule call inside the instrument. This can be protected by a control-flow statement (if ... then) to launch exactly n instances. We can use one of the instrument parameters (e.g. the last) to carry an instance count, and pass the others unchanged, as they are shared by all instrument instances.

At this point, we can define what the instrument needs as input parameters. An overall amplitude scaling and fundamental frequency are two basic parameters. We will also need to set the minimum and maximum band centre frequencies, the width of each band, and the total number of them. In order to space these apart in a perceptually even way we will use the following exponential interpolation expression to get the nth centre frequency f_n:

$$f_n = f_{min} \times \left(\frac{f_{min}}{f_{max}} \right)^{n/N} \tag{4.48}$$

where N is the total number of bands, and f_{min} and f_{max} are the minimum and maximum frequencies, respectively. To match this, the width of each band should be set by a Q factor (from which a bandwidth is derived).

Finally, the code can be completed by also alternatively tracking the fundamental frequency of the input sound (when that makes sense). To do this, we use one of the pitch trackers in Csound (in this case, `ptrack`) and smooth its result to remove any analysis jitter. We switch this tracking on if the fundamental is set to 0:

```
if ifo == 0 then
  kfo,ka ptrack asig,1024
  kfo  = port(kfo,0.01)
else
  kfo = ifo
endif
```

The full instrument is shown in Listing 4.23.

Listing 4.23 The ModFM vocoder instrument.

```
ksmps = 64
instr 1
 iscl = p4
 ifo  = p5
 iq   = p6
 imn  = p7
 imx  = p8
 ibnd = p9
 icnt = p10

 ifc = imn*(imx/imn)^(icnt/ibnd)
 ibw = ifc/iq

 asig in

 if ifo == 0 then
  kfo,ka ptrack asig,1024
  kfo  = port(kfo,0.05)
 else
  kfo = ifo
 endif

 asig = butterbp(asig,ifc,ibw)
 asig = butterbp(asig,ifc,ibw)
 krms = rms(asig)

 icor = 4.*exp(-1)
 kg   = exp(-kfo/(.29*ibw*icor))
 kg   = 2*sqrt(kg)/(1.-kg)
```

```
knd = kg*kg/2.
kfr  = ifc/kfo
kfn  = int(kfr)
kam  = kfr - kfn
aphs = phasor(kfo)
asin = table:a(aphs,-1,1,0,1)
aexp = exp(knd*asin - knd)
acr1 = tablei:a(aphs*kfn,-1,1,0.25,1)
acr2 = tablei:a(aphs*(kfn+1),-1,1,0.25,1)
asig = (kam*acr2+(1-kam)*acr1)*aexp
    out asig*krms*iscl

if icnt < ibnd then
  schedule(1,0,p3,p4,p5,p6,p7,p8,p9,icnt+1)
 endif
endin
```

This instrument takes its input (for analysis) from `in`, which is connected to the input buffer in Csound (just as `out` is connected to the output). This means that if we want to use it, we will need to set the option `-i. . . .` This can point to the computer soundcard (ADC) (`-iadc`) or to other sources, such as a soundfile, for instance, by passing its name (e.g. `-isound.wav`) to the option[9]. We also set `ksmps` to 64, which is a recommended setting for full-duplex audio (using input and output).

Typically, the instrument could be run with the following parameters:

```
schedule(1,0,15,0.7,0,5,80,8000,24,0)
```

tracking the fundamental, and

```
ifun = cpspch(7.00)
schedule(1,0,15,0.7,ifun,5,ifun,8000,24,0)
```

with a fixed pitch. When we are tracking the fundamental, some adjustments to the pitch-tracking part might be needed to fine-tune it to the input signal.

4.10.3 Resonant synthesis

One of the interesting aspects of summation formulae is how these methods pack signal generation and spectral modification into one single algorithm. We have already hinted at this in the previous examples and, in this section, we will study an instrument that mimics a source-filter combination with resonance control. As with the previous example, we could choose a number of summation formula techniques, as there is significant interchangeability between the methods (although each ones

---

[9] If we are using Python as a host, this option can be set with `cs.setOption("-iadc")` as per Listings 2.1 and 2.2.

has its specific characteristics). Our choice will be to use the generalised formulae, for both single- and double-sided spectra. We will employ the non-band-limited forms as they are simpler and work well if handled with care.

Starting with the single-sided formula, we have seen that we can reproduce the action of a low-pass filter by changing the a parameter, $0 \leq a < 1$. When this gets close to 1, we have a wide-band spectrum. On the other hand, when a = 0, we have a sinusoid. For example:

```
instr 1
  iamp = p4
  ifo = p5
  ka = line(0,p3,0.9)
  kasq = ka*ka
  aph  = phasor(ifo)
  asw  = tablei:a(aph,-1,1,0,1)
  aco  = tablei:a(aph,-1,1,0.25,1)
  asb = asw/(1 - 2*ka*aco + kasq)
  kscal = sqrt(1 - kasq)
  kamp = linen(kscal*iamp,0.1,p3,0.1)
     out(asb*kamp)
endin
```

```
schedule(1,0,2,0dbfs/4,200)
```

In this case, we have the equivalent of a non-resonant low-pass filter. To add a resonance peak, we can use the double-sided formula, which can emulate a bandpass source-filter arrangement. In this case, we choose one of our frequencies to be the centre of resonance, and another to be the fundamental frequency (as in Sect. 4.10.2). In order for the centre frequency to move, as before, we use two 'carriers':

```
instr 1
  iamp = p4
  ifo = p5
  kfc = line(p6,p3,p6*2)
  ia = (1-p7)
  iatt = p8

  iasq = ia*ia

  kfr  = kfc/ifo
  kfn  = int(kfr)
  kam  = kfr - kfn

  aph  = phasor(ifo)
  aco  = tablei:a(aph,-1,1,0.25,1)
  asw1 = tablei:a(aph*kfn,-1,1,0,1)
  asw2 = tablei:a(aph*(kfn+1),-1,1,0,1)
```

```
aden = (1 - 2*ia*aco + iasq)
adb1 = (1-iasq)*asw1/aden
adb2 = (1-iasq)*asw2/aden
adb =  (1-kam)*adb1 + kam*adb2

isc = sqrt((1-iasq)/(1+iasq))
asig =  adb/isc

    out(asig*iamp)
endin
```

```
schedule(1,0,2,0dbfs/4,200,2000,0.7,0.01)
```

Now we can mix these two into a single instrument, which will model a wide-band source shaped by a resonant filter. We use a parameter to control the amount of resonance (p7), which determines the a value for the band-pass component and is used to determine a mix between this and the low-pass signal. With more resonance, there will be less of the latter and more of the former, and vice versa. The bandwidth of the low-pass section is also controlled by the resonant frequency (p6), acting somewhat like a cut-off parameter. The full instrument is shown in listing 4.24. In Fig. 4.25, we can see plots obtained for four different levels of resonance using this instrument, with $f_0 = 200$ and $f_c = 3000$.

Listing 4.24 Resonant synthesis instrument.

```
instr 1
 iamp = p4
 ifo = p5
 ifc = p6
 ia2 = (1-p7)
 iatt = p8

 kfc  = expseg(ifc,iatt,ifc*4,p3-iatt,ifc)
 kfc = kfc*8 < sr ? kfc : sr/8;
 kN = int(8*kfc/(2*ifo)) - 1
 ka =  10^(-3/kN)
 kasq = ka*ka
 iasq2 = ia2*ia2

 aph  = phasor(ifo)
 asw  = tablei:a(aph,-1,1,0,1)
 aco  = tablei:a(aph,-1,1,0.25,1)
 asb = asw/(1 - 2*ka*aco + kasq)

 kfr  = kfc/ifo
 kfn  = int(kfr)
 kam  = kfr - kfn
```

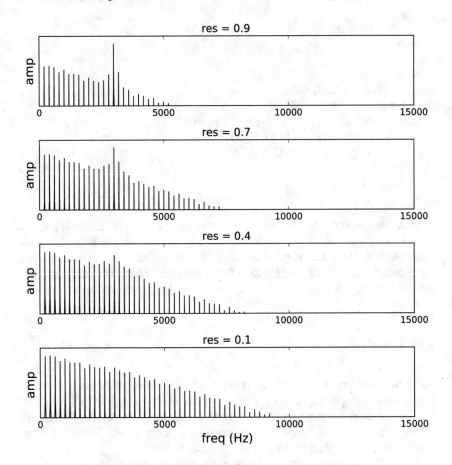

Fig. 4.25 Spectra for four different amounts of resonance in the instr of Listing 4.24.

```
asw1 = tablei:a(aph*kfn,-1,1,0,1)
asw2 = tablei:a(aph*(kfn+1),-1,1,0,1)
aden = (1 - 2*ia2*aco + iasq2)
adb1 = (1-iasq2)*asw1/aden
adb2 = (1-iasq2)*asw2/aden
adb =  (1-kam)*adb1 + kam*adb2

ksc1 = sqrt(1-kasq)
isc2 = sqrt((1-iasq2)/(1+iasq2))
amix = (1-p7)*asb/ksc1 + p7*adb/isc2
asig = linenr(amix*iamp/6,iatt,0.1,0.01)
```

```
     out(asig)
endin

icnt =0
while icnt < 1000 do
  schedule(1,icnt*0.125,.4,
         0dbfs/8,130*(1+abs(gauss(2)))),
         1000+(icnt*50)%1000,
         0.2+(icnt%10)/15,
         0.01)
  icnt += 1
od
```

This instrument can be driven suitably by a sequence as in the example given, or by other performance-related means that we will see later in this book. It demonstrates well the idea that summation formulae can replace the usual source-filter combination and that we are able to define a complex output spectrum with just a few parameters.

4.11 Conclusions

The techniques explored in this chapter provide a wide range of possibilities for instrument development. In many cases, they are close enough to each other that they can be used interchangeably in similar arrangements, with slightly different results. There is enough scope for a variety of uses, some of which were examined in detail in the three final instrument design case studies.

Some of these methods can be used in conjunction with other approaches. In particular, we might find that they can be used as source generators for further shaping by filters and other processes. Also, we will see in the next chapter that it is possible to extend some of these techniques in adaptive applications, where arbitrary input signals can be employed as control and modulation sources.

Chapter 5
Feedback and Adaptive Systems

Abstract This chapter explores a number of novel areas in sound synthesis and processing, where adaptive techniques are developed together with non-traditional uses of feedback systems. The text starts by discussing some applications of fixed feedback delay lines and the concept of waveguides. It moves on to look at pitch effects arising from variable delays, with which vibrato can be applied to arbitrary signals. This is then extended to audio-rate FM, giving rise to adaptive forms of FM/PM and related distortion techniques. The second part of the chapter takes a closer look at various forms of feedback modulation of amplitude and frequency, as well as the principles of periodic time-varying filters. Code examples in Csound and Faust illustrate each technique discussed.

The techniques explored in previous chapters can be seen as classic methods of sound synthesis, even though some of them have been developed quite recently. In this chapter, we will take a look at some more novel methods that explore principles of digital signal processing in a less traditional way. The ideas discussed in this chapter will be supported by a more detailed theory, which can be used as a reference for further studies and research in sound synthesis and processing.

5.1 Delay Lines

We have already seen the concept of signal delays in Chap. 3 and how these are used in the implementation of filters. In most cases, with the exception of FIR filters, the delays employed are of only a few samples. If we increase the length of a delay line, a number of new possibilities are opened up. For instance, once the time difference between an original sound and its delayed copy becomes of the order of around 50–80 ms, an echo effect will ensue. At a sampling frequency $f_s = 44100$, this can

© Springer International Publishing AG 2017
V. Lazzarini, *Computer Music Instruments*,
https://doi.org/10.1007/978-3-319-63504-0_5

be achieved with a delay line of around 2200–3500 samples[1], as shown in this code example:

Listing 5.1 Echo instrument.

```
ksmps = 64
instr 1
  a1 = in()
  a2 = delay(a1, 0.08)
    out(a1+a2)
endin
schedule(1,0,60)
```

These units can be used to create various echo and reverberation effects by combining various delayed signals, which could model the reflections in a given environment [22, 44, 59].

5.1.1 Waveguides

For the purpose of synthesis, we can use a delay line to model waves as they travel in the air inside a pipe or on a string [17]. For this, we need to feed the output of the delay line back into its input. This replicates the reflections at each end of the support structure (pipe or string), called a *waveguide*. In Csound, we can break down the delay function into two pieces to do the reading and writing separately. This allows us to create a feedback path, as demonstrated in the simple instrument in Listing 5.2.

Listing 5.2 Basic waveguide.

```
instr 1
  aex init p4
  a1 = delayr:a(1/p5)
      delayw(a1+aex)
    out(a1)
  aex = 0
endin
schedule(1,0,5,0dbfs/2,441)
```

The `delayr` opcode reads from a delay line and takes the time in seconds as a parameter, and the `delayw` complements this by writing into the same delay line. In order to make a sound with this delay line, we have to put a click into it (otherwise, it would be silent). So we fill an audio variable `aexc` initially and feed it to the delay

---

[1] Delay lines are implemented as buffers that are accessed circularly. Reading and writing proceeds at the sampling rate, at each sampling period, one sample is read and one sample is written, and the read/write position is incremented. When the end of the memory block is reached, that position is reset to the beginning of the buffer [59].

for one performance cycle only (`ksmps` samples), zeroing it afterwards. This click will keep recirculating in the delay line making a sound wave, which we send to the output. The pitch is inversely proportional to the size of the delay, the fundamental frequency being the reciprocal of the delay period.

We have implemented a basic *digital waveguide*. This is a very simple computer model of what happens in a pipe closed at both ends, or on a string, without taking into consideration any mechanical losses involved in the physical process. A simple way to account for these losses is to place an attenuation factor (just below unity) in the feedback path, so that at every repetition, the signal is reduced by a certain amount:

```
delayw(a1*0.99+aex)
```

The feedback delay structure that we have employed in this instrument is also known as a *comb* filter. The name relates to the shape of the amplitude response of this system, which contains a series of peaks centred at multiples of the fundamental, or *natural*, frequency f_0, of the filter. This, as discussed above, is the reciprocal of the delay time (T_0, see Fig. 5.1). This means that the filter can resonate at these frequencies when the feedback gain is close to unity[2].

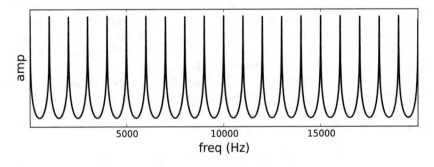

Fig. 5.1 Comb filter amplitude response, with delay time $T_0 = 1$ ms and natural frequency $f_0 = 1000$ Hz.

With a decay in place, the output starts to sound very much like a plucked string sound. Another way to model the losses is to consider the fact that the sound wave will lose high-frequency energy as it gets reflected. We can mimic this by using a simple FIR low-pass filter constructed out of the average of every two samples:

$$y(n) = \frac{x(n) + x(n-1)}{2} \tag{5.1}$$

---

[2] Pushing the feedback gain to one and above makes the filter unstable. We got away with this in the previous example (Listing 5.2) because we used a very short impulse as an input, but in the general case the gain has to be below unity.

where $y(n)$ and $x(n)$ are the inputs and outputs of this filter. In this case, the instrument would look like this:

Listing 5.3 Waveguide instrument with frequency-dependent loss.

```
instr 1
 aex init p4
 a1 = delayr:a(1/p5)
    delayw((a1+delay1(a1))/2+aex)
  out(a1)
 aex = 0
endin
schedule(1,0,5,0dbfs/2,441)
```

with the `delay1()` function delaying the signal by one sample. The result is even closer to a physical string. The model can be excited in different ways. We could initialise the delay line with a triangular wave that would resemble a string being plucked. The apex of the triangle in this case would relate to the plucking position. We could use an arbitrary audio signal as the excitation, which would make the instrument a model of a string resonator.

One remaining issue with this instrument is that it cannot be fine-tuned. Because delay lines are composed of an integral number of samples, their sizes can only increase or decrease by one sample, which means that the instrument can only be tuned to some pitches. The low-pass filter also adds a further 1/2-sample delay. To fix this, we have to find a means of adding a fractional delay to make the length of the whole digital waveguide match a certain fundamental.

Here we find an application for all-pass filters. These can be used to delay signals by fractional amounts without affecting the amplitude response of the signal. The type of all-pass filter we will use is a first order filter with a single coefficient that will be chosen to give the desired delay[59, 68]:

$$y(n) = c[x(n) - y(n-1)] + x(n-1) \tag{5.2}$$

where $x(n)$ is the input, $y(n)$ is the output, and c is determined from the fractional delay (in samples) d ($0 < d < 1$):

$$c = \frac{1-d}{1+d} \tag{5.3}$$

This filter, whose flow diagram is shown in Fig. 5.2 is implemented with the following Faust program lines:

```
c = (1-d)/(1+d);
process = _   +   (_ <: *(c) + _') ~ *(-c);
```

Now all we need to do is to calculate, based on the desired pitch, the size of the delay line and the required fractional delay. From the fundamental f_0, the truncated delay line size d_t is

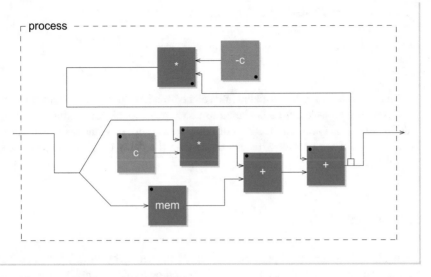

Fig. 5.2 First-order all-pass filter (Eq. 5.2).

$$d_t = \left\lfloor f_s/f_0 - \frac{1}{2} \right\rfloor \tag{5.4}$$

and the fractional delay d_f is:

$$d_f = f_s/f_0 - \left(d_t + \frac{1}{2}\right) \tag{5.5}$$

We can use these ideas to fix the original digital waveguide to allow fine tuning of the fundamental frequency. A Faust program implements the all-pass filter with a fractional delay and the Csound instrument uses it inside the delay line loop:

Listing 5.4 Well-tuned waveguide instrument.

```
giap faustcompile {{
d = hslider("del", 0, 0, 1, 0.00001);
c = (1-d)/(1+d);
process = _  +  (_ <: *(c) + _') ~ *(-c);
}}, "-vec -lv 1"

instr 1
 aex init p4
 idt = int(sr/p5 - 0.5)
 ifrc = sr/p5 - (idt + 0.5)
 a1 = delayr:a(idt/sr)
 alp = (a1+delay1(a1))/2
 ib,ap faustaudio giap,alp
```

```
faustctl(ib,"del",ifrc)
   delayw(ap+aex)
  out(a1)
 aex = 0
endin
```

This example shows how a delay line can be used in sound synthesis. More complex uses of feedback can be designed to accomplish sophisticated digital waveguide instruments. Other applications of delay lines (with or without feedback) can be developed if we provide a means of varying the delay time dynamically (and smoothly).

5.2 Variable Delays

There are a great number of possibilities that are opened up by allowing the delay to vary over time [20, 44]. By lengthening or shortening the delay line, it is possible to slide the pitch of a waveguide, for instance. Other effects can be created following this principle. The usual source of delay variation is a modulation signal, which can be produced, for instance, by a periodic waveform generator (oscillator, etc.; see Sect. 3.3.2).

The implementation of variable delays is usually done by employing a movable reading position, also known as a delay line tap. Writing proceeds as before, as in the fixed case, and samples are written one by one at the sampling rate, but now the reading can skip to different positions in the delay line memory rather than staying at a fixed time difference in relation to the write position. For a smooth result, we need to employ interpolation (just as we did in the case of oscillators), which can be linear, cubic or of a higher order.

5.2.1 Vibrato

An important side effect of varying the delay time is frequency modulation, which occurs when we read the delay memory at a different pace from that which we write to it. This difference in speed will make the pitch of the delay line contents to go up or down, depending on whether we are shortening the delay or lengthening it. This is exactly as if we had recorded an audio waveform at one rate and read it back at a different one, and so the delay memory can be considered as a kind of dynamic wavetable. Using variable delays, we can implement frequency modulation of arbitrary sounds.

As discussed above, the writing to the delay line proceeds at intervals of $\frac{1}{f_s}$ sec. To shorten the delay, the reading has to be faster than this, raising the playback pitch. If, however, we read it at a slower rate, lowering the pitch, the delay time is lengthened.

It is important to consider the *rate of change* of the delay time. If it is constant, and the lengthening or shortening happens at a constant pace, the speed of the reading will also be constant. In this case, the pitch will not vary, but will be higher or lower than the original. To achieve this, we can modulate the delay time with a straight-line waveform such as a triangle or ramp (sawtooth) signal.

On the other hand, if the rate of change of the delay line varies over time, the output pitch will also vary. This is the case for a non-linear modulating source, such as a sine wave. In this case, the read-out speed will be variable, resulting in a continuous pitch oscillation.

Both the amount of modulation and its rate affect the amount of pitch change in delay-line FM. Let's consider how these parameters contribute to the frequency deviation, depending on the modulation source. Since the key element is the rate of change of the delay, the resulting FM is actually the time derivative of the delay modulation. The effect will be dependent on the shape of the modulating waveform, as well as on its rate and amplitude (depth).

The derivative of the triangle waveform is a square wave. So for such a modulating signal with depth Δ_d seconds, and frequency f_m Hz, $\Delta_d \text{Tri}(f_m t)$, we have

$$\Delta_d \frac{\partial}{\partial t} \text{Tri}(f_m t) = \Delta_d f_m \text{Sq}(f_m t) \tag{5.6}$$

where $\text{Sq}(ft)$ is a square wave with frequency f.

In this case the frequency will alternate between two values, $f_0(1 \pm \Delta_d f_m)$, where f_0 is the fundamental frequency of the original input sound. For example, for a triangular oscillator with $f_m = 1$ Hz, and a delay between 0.1 and 0.3 sec ($\Delta_d = 0.2$), the pitch will jump between two values, $0.8 f_0$ and $1.2 f_0$.

Now we consider a non-linear modulator, such as a cosine wave, whose derivative is a sine. If the delay modulation signal is $0.5 \cos(2\pi f_m t)$ (scaled to fit the correct delay range, which also needs to be non-negative), we have

$$\Delta_d \frac{\partial}{\partial t} 0.5 \cos(2\pi f_m t) = -\Delta_d \pi f_m \sin(2\pi f_m t) \tag{5.7}$$

In this case the frequency will vary in the range $f_0(1 \pm \Delta_d \pi f_m)$, smoothly (rather than in jumps). An example instrument that produces a vibrato effect with arbitrary inputs is shown in Listing 5.5.

Listing 5.5 Vibrato instrument

```
ksmps = 64
instr 1
 asig = in()
 id = 0.5*p4/($M_PI*p5)
 ad = oscili(id,p5)
 ad += (id + 1/kr)
 a1 = delayr:a(1)
 av = deltapi(ad)
    delayw(asig)
```

```
  out (av)
endin
schedule (1,0,60,0.05,5)
```

In this example, p4 is the vibrato width interval w and p5 is the rate f_m. The amount of delay modulation is calculated according to the principles outlined in Eq. 5.7. The delay is set to range between $\frac{1}{kr}$ and the maximum delay ($\frac{1}{kr} + \frac{w}{\pi f_m}$). The reason for setting a minimum delay above 0 is to allow one extra position in the delay line for the interpolation operation, without which we would have artefacts (clicks) in the output.

5.2.2 Adaptive FM

While the examples above were considered from the point of view of low-frequency FM (or vibrato), there is nothing to stop us developing the idea for audio-rate FM[3], by a technique called *Adaptive FM* (AdFM) [57]. To develop the idea, let's first consider an input signal to the delay line consisting of a sinusoidal wave with frequency $\omega = 2\pi f_c$:

$$x(t) = \sin(\omega t) \tag{5.8}$$

and an arbitrary delay line modulation source, such as

$$m(t) = d_{max}D(t) \tag{5.9}$$

where d_{max} is the maximum delay and $D(t)$ is in the range $[0,1]$. The delay-line modulation of input signal can be defined as

$$y(t) = \sin(\omega[t - d_{max}D(t)]) \tag{5.10}$$

In this case the instantaneous frequency $\omega_i(t)$ of this modulated signal is the derivative of the sine function phase:

$$\omega_i(t) = \frac{\partial \omega[t - d_{max}D(t)]}{\partial t} = \omega - \frac{\partial D(t)}{\partial t}d_{max}\omega \tag{5.11}$$

from which we obtain the instantaneous frequency $f_i(t)$ of the output signal in Hz as

$$f_i(t) = f_c - \frac{\partial D(t)}{\partial t}d_{max}f_c \tag{5.12}$$

This provides the basis for the result seen in Eq. 5.7, in the case of a raised cosine modulator ($D(t) = 0.5 + 0.5\cos(\omega_m t)$), where $f_i(t)$ becomes

---

[3] Or, more accurately, PM.

$$f_i(t) = f_c + d_{max}\pi f_c f_m \sin(2\pi f_m t) \tag{5.13}$$

and from this we have the maximum frequency deviation $f_{max} = d_{max}\pi f_m f_c$. Putting this in terms of FM synthesis theory, we define the index of modulation I as the ratio $f_{max} : f_m$, and, as before, we define the delay time modulation width as $\Delta_d = d_{max} - d_{min}$ to replace d_{max} in the case of a minimum delay above zero, which yields

$$\Delta_d = \frac{I}{\pi f_c} \tag{5.14}$$

This puts the delay modulation in terms of the standard FM parameters I and f_c, which allows us to apply the synthesis equations given in Chap. 4 directly here.

The significance of applying delay-line modulation in FM synthesis can be realised when we use arbitrary inputs in place of the sinusoidal carrier signal in Eq. 5.8. Let's consider an input to the delay line with N partials of amplitudes a_n, frequencies ω_n and phases θ_n:

$$x(t) = \sum_{n=0}^{N-1} a_n \sin(\omega_n t + \theta_n) \tag{5.15}$$

The resulting FM (or PM), obtained through delay time modulation, is now

$$x(t) = \sum_{n=0}^{N-1} a_n \sin(\omega_n t + I_n \sin(\omega_m t) + \theta_n) \tag{5.16}$$

whose expansion into a sum becomes:

$$\sum_{n=0}^{N-1} a_n \left[\sum_{k=-\infty}^{\infty} J_k(I_n) \sin(\omega_n t + k\omega_m t + \theta_n) \right] \tag{5.17}$$

Each individual I_n is equal to $\Delta_d \pi f_n$ and thus we can derive them from the index I (Eq. 5.14):

$$I_n = I \frac{f_n}{f_c} \tag{5.18}$$

The result is that higher partials are modulated more intensely relative to lower ones, resulting in a bright spectrum for small amounts of modulation.

In order for us to be able to set a given $c : m$ ratio and index I for this algorithm, we need to extract the fundamental frequency of the input sound. This is the adaptive side of the technique. Once have we tracked the pitch, we can determine the modulation parameters from it. This is demonstrated in Listing 5.6, where we use a flute tone as input. This particular sound is tuned to C (pch = 9.00), and we use this as the basis for transposing the sound to other pitches (using the opcode diskin)[4].

---

[4] Since we know the pitch of this sound, we could have dispensed with pitch tracking in this case. However, we are using pitch tracking here to give the general form of the algorithm, which does not depend on prior knowledge of the characteristics of the input.

Listing 5.6 AdFM example.

```
instr 1
 irt = p5
 as  = diskin:a("flutec3.wav",p4/cpspch(9.00))
 kcps,kamp ptrack as,512
 anx = linen(p6,4,p3,2)
 adt = oscili(anx/($M_PI*kcps),kcps/irt,-1,0.25)
 adt = 0.5 + adt*0.5 + 1/kr
 adp  = delayr:a(1)
 adel = deltap3(adt)
    delayw (as)
       out(adel)
endin
schedule(1,0,filelen("flutec3.wav"),440,1,3)
```

In this example, the parameters are the (carrier) pitch, $c : m$ ratio and index of modulation. A plot of the original flute spectrum and the AdFM output is shown in Fig. 5.3.

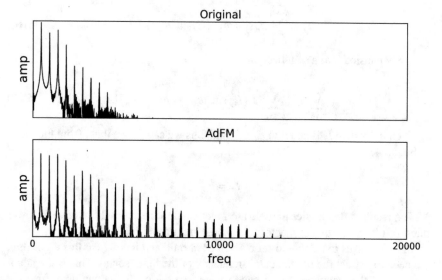

Fig. 5.3 Original and adaptive FM spectra of a flute tone.

5.2.3 Self-modulation

A related technique is that of self-modulation [83], a type of complex-carrier complex-modulator PM where the input signal is used for both functions. In this case, we have to make sure the signal is within the normalised $[-1,1]$ range, ideally full-scale. This is then used to replace the modulator in Listing 5.6. We can still track the fundamental frequency as before, as a reference for setting the modulation width from an index of modulation. If the signal is not full scale, a larger index value might be needed to provide a significant effect.

Listing 5.7 Self-modulation example.

```
instr 1
 irt = p5
 as  = diskin:a("flutec3.wav",p4/cpspch(9.00))
 kcps,kamp ptrack as,512
 anx = linen(p6,4,p3,2)
 adt = (as/0dbfs)*anx/($M_PI*kcps)
 adt = 0.5 + adt*0.5 + 1/kr
 adp = delayr:a(1)
 adel= deltap3(adt)
    delayw(as)
       out(adel)
endin
schedule(1,0,filelen("flutec3.wav"),440,1,8)
```

Figure 5.4 illustrates this effect. Note that this variation of AdFM is not limited to monophonic periodic input waveforms, as there is no need to set a modulator frequency according to a given pitch.

5.2.4 Asymmetrical methods

As the AdFM technique is basically an alternative implementation of the PM/FM algorithm, it is possible to modify it to create variations, such as the ones we saw in Chap. 4. The first of these follows from the realisation that the upper and lower sidebands of FM can be obtained separately by use of the formula in Eq. 4.29, which is based on ring-modulating the output of simple FM with an exponential wave-shaping of a cosine waveform [69]:

$$\exp(I\cos(\omega_m) - I)\cos(\omega_c + sI\sin(\omega_m)) \qquad (5.19)$$

The output will contain only the upper or the lower sidebands, depending on the value of s. If this is -1, then components below the carrier frequency will be produced. If it is 1, then the spectrum will feature only the upper sidebands. This

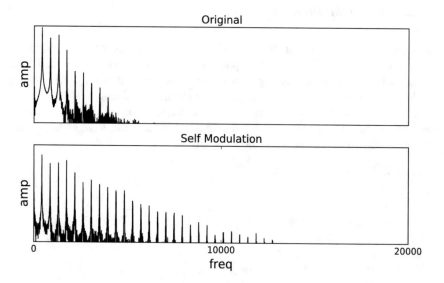

Fig. 5.4 Original and self-modulation spectra of a flute tone.

single-sideband expression[5] can be applied to delay-based FM with the addition of ring modulation [56]. In order to produce the desired effect of sideband cancellation, it is important that phases are correctly aligned. Carrier phases are not relevant in this case, as we need only to match the two modulator phases.

In order to match the phase offsets in Eq. 5.19, the variable delay line modulator needs to be given an offset of $\frac{\pi}{2}$ radians (relative to cosine phase, making it an inverted sine wave), and to be scaled by s to provide either the upper or the lower side of the spectrum. The wave-shaped sinusoid is in cosine phase. The modified instrument is shown in Listing 5.8 and includes one extra parameter (s), set by p7.

Listing 5.8 Single-sideband adaptive FM example.

```
instr 1
 irt = p5
 is = p7
 as  = diskin:a("flutec3.wav",p4/cpspch(9.00))
 kcps,kamp ptrack as,512
 anx = linen(p6,4,p3,2)
 adt = oscili(-is*anx/($M_PI*kcps),kcps/irt)
 adt = 0.5 + adt*0.5 + 1/kr
 adp = delayr:a(1)
 adel = deltap3(adt)
    delayw (as)
```

---

[5] This is a special case of the extended ModFM formula discussed in Chap. 4.

```
amod = oscili(anx,kcps/irt,-1,0.25)
       out(adel*exp(amod-anx))
endin
schedule(1,0,filelen("flutec3.wav"),440,1,3,1)
```

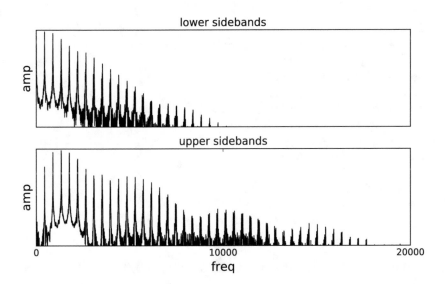

Fig. 5.5 Single-sideband adaptive FM spectra of a flute tone (separate lower and upper sidebands).

The spectra obtained from lower and upper-sideband AdFM are shown in Fig. 5.5. Notice the difference between these plots and Fig. 5.3, which uses the original double-sided formula.

The Asymmetrical FM algorithm (Eq. 4.15) is actually a generalisation of the principles in this single-sideband formula. It includes the introduction of an extra parameter, r, which controls the spectral symmetry of the FM spectrum:

$$a_s \exp\left(\frac{k}{2}(r - \frac{1}{r})\cos(\omega_m)\right) \times \sin\left(\omega_c + \frac{k}{2}(r + \frac{1}{r})\sin(\omega_m)\right) \quad (5.20)$$

The application of Eq. 5.20 to delay-based FM is straightforward [56]. With careful setting of the modulator phases, as before, we now require only the use of an expression that includes the parameter r. For normalisation purposes, a_s is set to

$$a_s = \exp\left(-\frac{1}{2}\log I_0\left(k\left[r - \frac{1}{r}\right]\right)\right) \quad (5.21)$$

in which the logarithm of the modified Bessel function of order 0 (I_0) can be obtained by table lookup. Csound has a function generator (GEN 12), which can create a table for this. Listing 5.9 shows an example of the application of this algorithm to

AdFM. The function table is generated by `ftgen` and the lookup is performed by `tablei`.

Listing 5.9 Asymmetrical adaptive FM example.

```
gifn = ftgen(0,0,16384,-12,20)
instr 1
 irt = p5
 ir = p7
 as  = diskin:a("flutec3.wav",p4/cpspch(9.00))
 kcps,kamp ptrack as,512
 anx = linen(p6,4,p3,2)
 anx1 = anx/2*(ir + 1/ir)
 adt = oscili(anx1/($M_PI*kcps),kcps/irt)
 adt = 0.5 + adt*0.5 + 1/kr
 adp = delayr:a(1)
 adel = deltap3(adt)
    delayw (as)
 anx2 = anx/2*(ir - 1/ir)
 amod = oscili(.5*anx2,kcps/irt,-1,0.25)
 aln  = tablei:a(anx2,gifn)
        out(adel*exp(amod+aln))
endin
schedule(1,0,filelen("flutec3.wav"),440,1,3,2)
```

We can observe the result of varying the parameter r in Fig. 5.6, where it has been used to shift the spectral peak upwards or downwards ($r > 1$ or $r < 1$, respectively).

5.2.5 Adaptive ModFM

Finally, using similar principles, the ModFM algorithm can easily be implemented for adaptive applications, by the substitution of the cosine wave carrier oscillator term in Eq. 4.27 by an arbitrary input signal $s(t)$ [53]:

$$\exp(k\cos(\omega_m) - k)s(t) \tag{5.22}$$

The timbral character of this effect is related to the other adaptive FM implementations, with some special characteristics due to the spectral qualities of ModFM discussed in Chap. 4, and is achieved without the need for a delay line. It has been observed that by setting $c : m$ to an integer above 1, we can obtain good-quality octave transposition. Also, an interesting result is that, with some settings, each harmonic of the input signal has a small formant region associated with it, with a distinct timbral signature. Two spectra obtained from adaptive ModFM are shown in Fig. 5.7, with $c : m$ ratios of 2 and $\frac{1}{4}$, where we can see clearly the octave transposition in the first case:

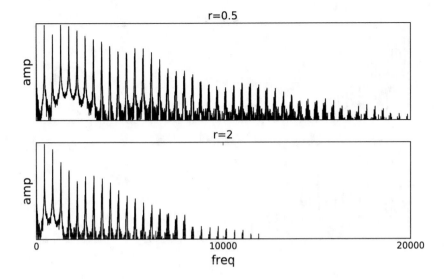

Fig. 5.6 Asymmetrical adaptive FM spectra of a flute tone (for two values of the symmetry parameter r).

Listing 5.10 Adaptive Modified FM example.

```
instr 1
 irt = p5
 as  = diskin:a("flutec3.wav",p4/cpspch(9.00))
 kcps,kamp ptrack as,512
 anx = linen(p6,4,p3,2)
 amod = oscili(anx,kcps/irt,-1,0.25)
       out(as*exp(amod-anx))
endin
schedule(1,0,filelen("flutec3.wav"),440,1/4,2)
```

By bringing the delay line modulation into the process, we can also implement the extended ModFM algorithm of Eq. 4.28. The delay line modulator is an inverted sine, tuned to f_m, which provides an equivalent phase modulation effect, as discussed in the previous section. Listing 5.11 demonstrates this, with the application of the two extra parameters (r and s) in p7 and p8, respectively.

Listing 5.11 Extended adaptive Modified FM example.

```
instr 1
 irt = p5
 ir = p7
 is = p8
 as  = diskin:a("flutec3.wav",p4/cpspch(9.00))
```

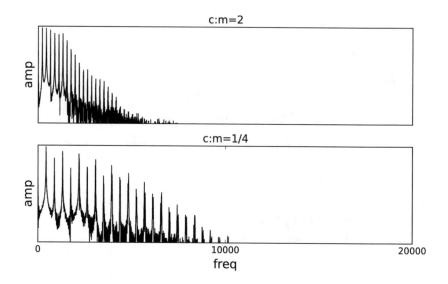

Fig. 5.7 Adaptive ModFM spectra of a flute tone (for two values of the $c : m$ ratio).

```
kcps,kamp ptrack as,512
anx = linen(p6,4,p3,2)
adt = oscili(is*anx/($M_PI*kcps),kcps/irt)
adt = 0.5 + adt*0.5 + 1/kr
adp = delayr:a(1)
adel = deltap3(adt)
    delayw(as)
anx2 = anx/2*(ir - 1/ir)
amod = oscili(ir*anx,kcps/irt,-1,0.25)
        out(adel*exp(amod-ir*anx))
endin
schedule(1,0,filelen("flutec3.wav"),440,1,5,1,0.5)
```

5.3 Heterodyning

As hinted at in the implementation of adaptive ModFM, it is possible to realise
a form of phase modulation that does not require delay-line modulation. This can
be done by transforming the relevant closed-form summation formula into a heterodyning arrangement, where two signals are combined through ring modulation.
Beginning with such an expression [57],

$$\sin(\omega_c)\cos(I sin(\omega_m)) = \frac{1}{2}[\sin(\omega_c + I\sin(\omega_m)) + \sin(\omega_c - I\sin(\omega_m))] \quad (5.23)$$

we can see that we have a combination of two PM signals as a result. This will lead to the cancellation of some components in the output signal, namely the odd sidebands. However, we can see that this formulation of PM allows us to implement an adaptive form by replacing the sine wave carrier. This provides a variation on the original AdFM that is worth exploring. If we employ an input signal such as that in Eq. 5.15, the synthesis output will be

$$\frac{1}{2}\sum_{n=0}^{N-1} a_n[\sin(\omega_n t + I\sin(\omega_m t) + \theta_n) + \sin(\omega_n t - I\sin(\omega_m t) + \theta_n)] \quad (5.24)$$

This has two important differences from the delay-line method: the cancellation of odd sidebands and a single invariant index of modulation (I) across all carriers. The result is a more gentle spectral modification, especially in terms of high harmonics, as demonstrated by Fig. 5.8. The implementation of this method is shown in Listing 5.12

Listing 5.12 Heterodyne adaptive FM.

```
instr 1
  irt = p5
  as  = diskin:a("flutec3.wav",p4/cpspch(9.00))
  kcps,kamp ptrack as,512
  anx = linen(p6,4,p3,2)
  amod = oscili(anx/($M_PI*2),kcps/irt)
        out(as*tablei:a(amod,-1,1,0.25,1))
endin
schedule(1,0,filelen("flutec3.wav"),440,1,5)
```

5.3.1 Adaptive PD

The principles outlined above can be extended further to provide an adaptive implementation of phase distortion (PD) [47]. As shown in Sect. 4.9, we can define the method with the following expression:

$$s(t) = -\cos(2\pi\phi_{pd}(\phi(t))) \quad (5.25)$$

which can be transformed into a PM expression by breaking down the phase shaping function $\phi_{pd}(x)$ into linear phase and modulation terms $x + \upsilon(x)$:

$$s(t) = -\cos(2\pi[\phi(t) + \upsilon(\phi(t))]) \quad (5.26)$$

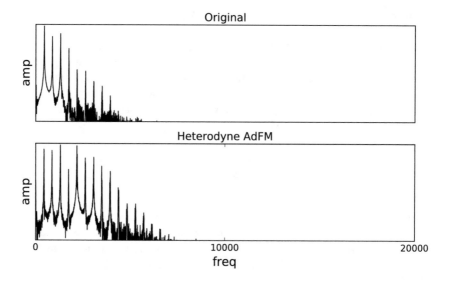

Fig. 5.8 Original and heterodyne AdFM spectra of a flute tone.

This expression can now be reworked as a heterodyne expression involving two terms:

$$-\cos(2\pi[\phi(t) + \upsilon(\phi(t))]) =$$
$$\sin(2\pi\phi(t))\sin(2\pi\upsilon(\phi(t))) - \cos(2\pi\phi(t))\cos(2\pi\upsilon(\phi(t)))$$

(5.27)

As before, we can separate the carrier signal cleanly from the modulator, allowing us to substitute the former by any arbitrary input. Through its reinterpretation as PM, PD is now implemented as a heterodyned form of wave shaping, using sinusoidal transfer functions. It is important that both the carrier and the modulator in the two terms have the correct phase offsets (relative to each other) to reproduce the output faithfully. To replace the carrier signal with an arbitrary input, we need only to preserve the 90-degree offset between the carriers in the two terms of Eq. 5.27. Given that these make up an analytic signal, we can use a Hilbert transform to place the input signal in quadrature.

The implementation of this algorithm is presented in Listing 5.13, and a plot of its output spectra with a flute tone input is shown in Fig. 5.9. If we are using a one-inflection PD function (as in Eq. 4.44, we can cast the modulation term $\upsilon(x)$ as a piecewise triangular function with two segments

$$v(x) = \begin{cases} \left(\frac{1}{2}-d\right)\frac{x}{d}, & x < d \\[2mm] \left(\frac{1}{2}-d\right)\left[1+\frac{(1-x)}{(1-d)}\right], & x \geq d \end{cases} \tag{5.28}$$

which is a sawtooth wave inflected at d with an amplitude of $0.5 - d$. This can be translated into a function table of length l built with GEN 7 in Csound, with two segments (ld and $l(1-d)$). As this is PM, we can also generate a timbral change by placing an envelope in the modulation oscillator.

Listing 5.13 Adaptive PD example.

```
isiz = 16384
gid = 0.1
gifn = ftgen(0,0,isiz,7,0,isiz*gid,1,isiz*(1-gid),0)
instr 1
 as  = diskin:a("flutec3.wav",p4/cpspch(9.00))
 kcps,kamp ptrack as,512
 knv = linen(0.5-gid,4,p3,1)
 amod = oscili(knv,kcps,gifn)
 asin = sin(2*$M_PI*amod)
 acos = cos(2*$M_PI*amod)
 ai,ar hilbert as
    out(asin*ai - acos*ar)
endin
schedule(1,0,filelen("flutec3.wav"),440)
```

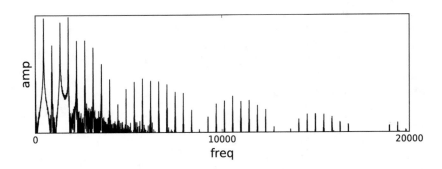

Fig. 5.9 Heterodyne adaptive PD spectrum of a flute tone input.

5.3.2 Adaptive SpSB

Similarly, split sideband synthesis with arbitrary input signals can be implemented using these ideas [56, 58]. To do so, we replace the sinusoidal carrier in Eq. 4.26, with an input signal that is placed in quadrature with a Hilbert transform filter. With an input $x(t)$, the algorithm becomes

$$
\begin{bmatrix} s_{upper,even}(t) & s_{upper,odd}(t) \\ s_{lower,even}(t) & s_{lower,odd}(t) \end{bmatrix} = \begin{bmatrix} x(t) & -H(x(t)) \\ x(t) & H(x(t)) \end{bmatrix} \times
$$
$$
\begin{bmatrix} \cos(I\sin(\omega_m)) & \sin(I\sin(\omega_m)) \\ H(\cos(I\sin(\omega_m))) & H(\sin(I\sin(\omega_m))) \end{bmatrix}
\tag{5.29}
$$

where, as before $H()$ represents the Hilbert transform operation, which delays the input by 1/4 of a cycle.

The implementation (Listing 5.14) follows this expression closely. Since we have four outputs, in this example we use a quadraphonic output by setting the number of channels (nchnls) to 4. The resulting spectra of the sideband groups are shown in Fig. 5.10.

Listing 5.14 Adaptive SpSB example.

```
nchnls=4
instr 1
 irt = p5
 as   = diskin:a("flutec3.wav",p4/cpspch(9.00))
 kcps,kamp ptrack as,512
 anx = linen(p6,4,p3,2)
 amod = oscili(anx/(2*$M_PI),kcps/irt)
 acos = tablei:a(amod,-1,1,0.25,1)
 asin = tablei:a(amod,-1,1,0,1)
 aae,abe hilbert acos
 aao,abo hilbert asin
 ac,ad hilbert as
 aeu = aae*ac - abe*ad
 aou = aao*ac - abo*ad
 ael = aae*ac + abe*ad
 aol = aao*ac + abo*ad
     out(aeu,aou,ael,aol)
endin
schedule(1,0,filelen("flutec3.wav"),440,1,2)
```

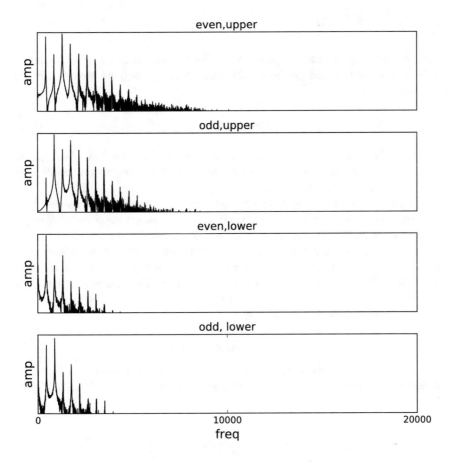

Fig. 5.10 Adaptive SpSB spectra of a flute tone input, showing the four separate sideband groups.

5.4 Feedback Modulation

The use of feedback can be extended to modulation applications, with which we can create a number of novel instrument designs. Such recursive signal paths can be applied to either amplitude or frequency modulation. In this section, we will examine these methods and introduce the related theory of periodic time-varying filters.

5.4.1 Feedback AM

The heterodyning operation explored in earlier sections of this chapter is actually a form of audio-rate AM synthesis [38]. This can, in general, be expressed by the following equation,

$$s(t) = (m(t) + a)\frac{c(t)}{b} \tag{5.30}$$

which involves a carrier signal $c(t)$ and a modulator $m(t)$. If $a = b \neq 0$, we have the standard formulation of AM. If $a = 0$, the expression implements ring modulation with $b \neq 0$ as an inverse scaling factor.

For sinusoidal signals, the results of AM are simple to analyse: the carrier signal plus two extra frequencies (one sideband) on each side, the sums and differences of the carrier and modulator. To produce a richer spectrum, we can either employ a component-rich modulator (as we did in the example of ring-modulated noise of Chap. 3), or, alternatively, employ a feedback path to provide the modulation. This idea, a feedback AM (FBAM) oscillator, was first implemented as one of the instruments in Risset's catalogue of computer-synthesised sounds [86].

The FBAM algorithm is defined by

$$y(t) = \cos(\omega_0 t)[1 + y(t-1)] \tag{5.31}$$

with a fundamental frequency $\omega_0 = 2\pi f_0$, and $y(t-1)$ representing a 1-sample delay of the output. The feedback expression can be expanded into a sum of products given by

$$
\begin{aligned}
&\cos(\omega_0 t) + \cos(\omega_0 t)\cos(\omega_0[t-1]) + \\
&\cos(\omega_0 t)\cos(\omega_0[t-1])\cos(\omega_0[t-2]) + \ldots \\
&= \sum_{k=0}^{\infty}\prod_{\tau=0}^{k}\cos(\omega_0[t-\tau])
\end{aligned} \tag{5.32}
$$

from which we deduce that the spectrum is composed of various harmonics of the fundamental, f_0. The output spectrum and waveform generated by the FBAM algorithm are shown in Fig. 5.11.

The complexity of the product in Eq. 5.32 makes it hard to perform any further analysis. Instead, we can, reinterpret FBAM as an IIR system. Equation 5.31 can be re-cast as a coefficient-modulated one-pole recursive filter whose input is a sinusoid:

$$y(t) = x(t) + a(t)y(t-1) \tag{5.33}$$

where we have $x(t) = a(t) = \cos(\omega_0 t)$.

This is a periodic linear time-varying (PLTV) system, which has a number of important characteristics. First, it does not have a single fixed impulse response, but a periodically time-varying one. Second, the spectral properties of this system are

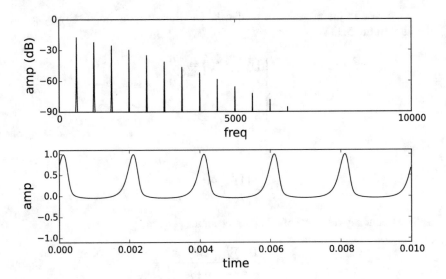

Fig. 5.11 FBAM waveform and spectrum.

based on their own functions of the discrete time. The output of the filter not only depends on the current coefficients, but also on the values they assumed in the past.

Our analysis starts with a non-recursive PLTV filter defined by

$$y(t) = \sum_{k=0}^{T} b_k(t) x(t-k) \tag{5.34}$$

where $b_k(t)$ are the time-varying coefficients of the delays $x(t-k)$. The impulse response of this filter is defined as $y(t)$ measured at time t in when an impulse $x(\tau) = \delta(\tau)$ is applied at time τ [14]:

$$h(\tau,t) = \sum_{k=0}^{T} b_k(t) \delta(t-k-\tau) \tag{5.35}$$

From this, we can obtain the frequency response of the filter[6] at time t as

$$H(\omega,t) = \sum_{k=0}^{T} b_k(t) e^{-jk\omega} \tag{5.36}$$

The impulse and frequency responses of a non-recursive PLTV for a given time point are fairly equivalent to that of an ordinary time-invariant system. This is not

---

[6] The frequency response $H(\omega)$ of a filter is a complex-valued function of frequency, which packs the amplitude and phase responses together. The former is the magnitude of the frequency response ($|H(\omega)|$) and the latter its angle ($\arg(H(\omega))$, see Appendix A (Sect. A.4). In the present case, the frequency response is also time-varying and so has an extra parameter t attached to it.

quite the case in the presence of feedback. The time-varying impulse response of the filter in Eq. 5.33 is:

$$h(\tau,t) = \begin{cases} \prod_{i=\tau+1}^{t} a(i) = \frac{g(t)}{g(\tau)}, & \tau < t \\ 1, & \tau = t \\ 0, & \tau > t \\ 0, & t < 0 \end{cases} \qquad (5.37)$$

where

$$g(t) = \begin{cases} \prod_{i=1}^{t} a(i), & t > 0 \\ 1, & t = 0 \end{cases} \qquad (5.38)$$

The frequency response of this filter becomes

$$H(\omega,t) = \frac{1 + \sum_{k=1}^{T-1} h(t-k,t) e^{-jk\omega}}{1 - g(T) e^{-jT\omega}} \qquad (5.39)$$

where T is the period (rounded to the nearest sample) of the coefficient-modulating signal ($a(t)$). This is the general case. If the modulator is a cosine wave whose period is $T_0 = \frac{1}{f_0}$, then we can set $T = \lfloor T_0 + 0.5 \rfloor$. In this case (Eq. 5.33), the frequency response turns out to be

$$H(\omega,t) = \frac{1 + \sum_{k=1}^{T-1} b_k(t) e^{-jk\omega}}{1 - a_T e^{-jT\omega}} \qquad (5.40)$$

where $b_k(t)$ is given by

$$b_k(t) = h(t-k,t) = \prod_{i=t-k+1}^{t} a(i) = \prod_{i=1}^{k} \cos(\omega_0[i+t-k]) \qquad (5.41)$$

and a_T is given by

$$a_T = g(T) = \prod_{i=1}^{T} a(i) = \prod_{i=1}^{T} \cos(\omega_0 i) \qquad (5.42)$$

This IIR system is equivalent to a filter of length T, made up of a cascade of a time-varying FIR filters of order $T - 1$ and coefficients $b_k(t)$, and a comb filter with a fixed coefficient a_T:

$$y(t) = x(t) + \sum_{k=1}^{T-1} b_k(t) x(t-k) + a_T y(t-T) \qquad (5.43)$$

The feedback section has an influence on the stability of the filter, which depends on the condition $|a_T| < 1$, which is the case for Eq. 5.31. The time-varying FIR section of this filter is responsible for the generation of harmonic partials and the overall spectral envelope of the output.

In order to make the algorithm more flexible, a means of controlling the amount of modulation (distortion) needs to be inserted into the system. This can be effected by introducing a modulation index β into Eq. 5.31:

$$y(t) = \cos(\omega_0 t)[1 + \beta y(t-1)] \qquad (5.44)$$

By varying the parameter β, we can produce dynamic spectra, from a sinusoid to a fully modulated signal with various harmonics. This demonstrated in Fig. 5.12, which shows a spectrogram of an FBAM signal with β varying linearly from 0 to 1.5. The output bandwidth and the amplitude of each partial increase with the β parameter. This is a simpler relation than that in FM synthesis, in which the partial amplitudes fluctuate with changes in the modulation index. The maximum value of β depends on the aliasing tolerance: higher values of β will increase the signal bandwidth significantly. The stability of the filter is also dependent on β, the condition now becoming

$$|\beta a_T| < 1 \qquad (5.45)$$

However, the stability of the system is not the limiting issue for β as aliasing will dominate the spectrum for values well within the stable range. For this reason, the amount of modulation needs to be controlled carefully in the case of high frequencies, the maximum varying from around 1.8 at 100 Hz to 0.3 at 4 kHz.

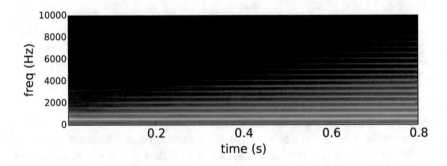

Fig. 5.12 Spectrogram of an FBAM tone with dynamic spectrum.

The gain of an FBAM system varies significantly with β in a frequency-dependent way. It also grows rapidly for β values beyond unity. Some means of normalisation is required, and the best method for this is to adaptively control the output based on an RMS measurement of the filter input. This can be by using, for instance, the `balance` unit in Csound, which measures the RMS values of the input and the output and applies gain scaling that is proportional to their ratio (see Fig. 5.13). RMS estimation is done by taking the absolute value of the signal and then low-pass filtering (using a tone-type first-order filter).

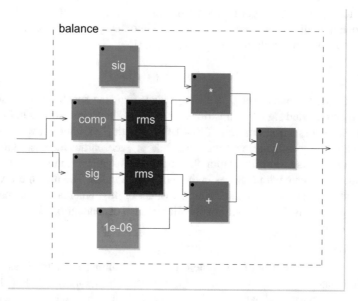

Fig. 5.13 RMS balance unit flowchart, where the comparator (comp) signal is used to control the amplitude of an input.

The main component for an implementation of FBAM is the PLTV filter of Eq. 5.33. This is a specialist component that is not normally built into music programming systems. For this reason, we will write a simple Faust program to implement it. The basic FBAM filter is given by

```
fbam(beta) = ((+(1))*(_)~*(beta));
```

whose flowchart is shown in Fig. 5.14.

The whole algorithm can be rendered in a complete Faust instrument by adding the driving sinusoid and the balancing unit, with sliders to control its parameters (Listing 5.15]. The Faust library function `osci` is used as the oscillator input.

Listing 5.15 FBAM instrument code.

```
import("music.lib");

vol = hslider("amp", 0.1, 0, 1, 0.001) : lp(10);
fr = hslider("freq [unit:Hz]", 440,110,880,1) : lp(10);
beta = hslider("beta", 1,0,1,0.001) : lp(10);

lp(fr) = *(1 - a1) : + ~ *(a1) with
    { C = 2. - cos(2.*PI*fr/SR);
        a1 = C - sqrt(C*C - 1.); };
rms = fabs : lp(10);
balance(sig, comp) = sig*rms(comp)/(rms(sig)+0.000001);
```

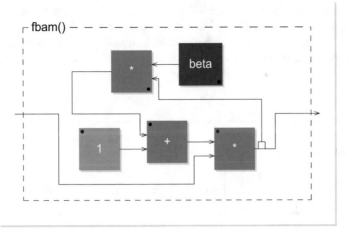

Fig. 5.14 FBAM PLTV filter flowchart.

```
fbam(beta) = ((+(1))*(_)~*(beta));
process = vol*(osci(fr) <: fbam(beta),_ : balance);
```

In this code, we see the `with` notation, which can be used to simplify the calculation of parameters such as filter coefficients. The general form of this is

```
function(...) = expression with
   { parameter computation };
```

While this does not add any new programming facilities to the language (it is what we call *syntactic sugar*), it allows for much more readable code. As an alternative to placing the whole algorithm as a self-contained instrument in Faust, we could have used it to implement just the PLTV filter part of FBAM, integrating it inyo a Csound instrument, as we did for waveguide tuning.

FBAM variations

FBAM provides a basic platform on which new algorithms can be built. A number of these will be examined here: the insertion of a feedforward term; an all-pass filter-like structure; other forms of heterodyning; and non-linear distortion [49]. Such variations inevitably lead to considering FBAM as a specific case of PTLV filters in general [38].

By including a feedforward delay term in the basic FBAM equation, it is possible to generate a different wave shape. In this case the expression becomes

$$y(t) = \cos(\omega_0[t-1]) - \cos(\omega_0 t)[1 + \beta y(t-1)] \tag{5.46}$$

However, apart from a change in the DC term, there is no other significant modification to the spectrum. The addition of the feedforward delay does not change the input, as it remains a sinusoid, but it introduces a half-sample delay. This is responsible for the change in the shape of the output waveform, which is different owing to different phase offsets in its harmonics (Fig. 5.15).

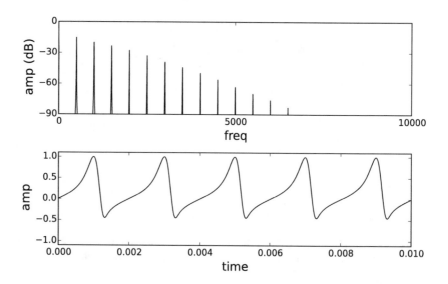

Fig. 5.15 Waveform and spectrum of FBAM variation 1.

Following the addition of the feedforward term, we can create a *coefficient-modulated* all-pass topology[7] filter. In this case, the general form of the filter is

$$y(t) = x(t-1) - a(t)[x(t) - y(t-1)] \tag{5.47}$$

which yields the following FBAM variation:

$$y(t) = \cos(\omega_0[t-1]) - \beta \cos(\omega_0 t)[\cos(\omega_0 t) - y(t-1)] \tag{5.48}$$

This can be equated to a form of PM synthesis [101]. This variant exhibits similar stability and aliasing behaviour to the basic FBAM algorithm, allowing us to raise β above unity for increased effect. The spectrum and waveform for this variation are shown in Fig. 5.16.

The Faust code for this variation is presented in Listing 5.16, and a flowchart for the corresponding PLTV filter is shown in Fig. 5.17.

---

[7] Since the behaviour of a PLTV filter is not strictly all-pass as it is in the time-invariant case, the term *all-pass topology* might be more suitable here.

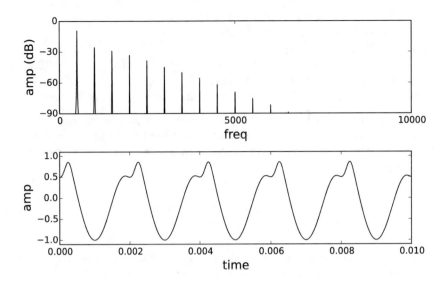

Fig. 5.16 Waveform and spectrum of FBAM variation 2.

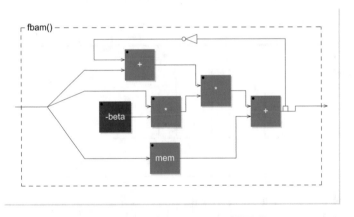

Fig. 5.17 Flowchart of FBAM variation 2 (all-pass topology) PLTV filter.

Listing 5.16 FBAM variation 2 instrument code.

```
import("music.lib");

vol= hslider("amp", 0.1, 0, 1, 0.001)  : lp(10);
freq = hslider("freq [unit:Hz]", 440,110,880,1) : lp(10);
beta = -1*hslider("beta", 1,0,1,0.001) : lp(10);

lp(freq) = *(1 - a1) : + ~ *(a1) with
```

```
{ C = 2. - cos(2.*PI*freq/SR);
    a1 = C - sqrt(C*C - 1.); };
rms = fabs : lp(10);
balance(sig, comp) = sig*rms(comp)/(rms(sig)+0.000001);
fbam(beta) = _ <: ((+(_))*(_*beta) + _' ~ _*(-1));
process = vol*(osci(freq) <: fbam(beta),_ : balance);
```

Another variation can be developed by adding an extra ring modulator to the formula. In this case, the original FBAM algorithm is used to create a signal with a baseband spectrum, which is then heterodyned by a cosine carrier with a frequency θ, as defined by

$$y(t) = \cos(\omega_0 t)[1 + \beta y(t-1)]$$
$$s(t) = \cos(\theta t)y(t) \tag{5.49}$$

This is very similar to some of the double-sided formulae seen in Chap. 4. The principle is very useful for generating resonant spectra and formants by setting θ to $k\omega_0$ (where k is an integer above 0). This makes the carrier frequency a multiple of the FBAM fundamental and the centre of a resonant region, as illustrated in Fig. 5.18. The parameter *beta* in this case becomes a bandwidth control.

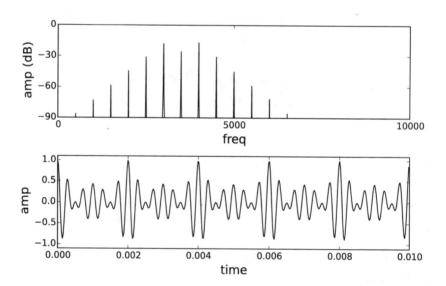

Fig. 5.18 Waveform and spectrum of FBAM variation 3, with $k = 7$, $f_0 = 500$ Hz and $\beta = 0.4$.

More generally, we can adjust this algorithm to create a formant region at any frequency by using two carriers that are set to adjacent harmonics around the centre of resonance (as we did in the phase-synchronous ModFM synthesis algorithm). In this case, we set k to

$$k = \left\lfloor \frac{f_c}{f_0} \right\rfloor \qquad (5.50)$$

and apply it to the two carriers

$$g = \frac{f_c}{f_0} - k$$
$$s(t) = y(t)[(1 - g)\cos(k\omega_0 t) + g\cos([k+1]\omega_0 t)] \qquad (5.51)$$

where $y(t)$ is the output of the FBAM expression (Eq. 5.44). A Faust program implementing this algorithm is shown in Listing 5.17.

Listing 5.17 FBAM variation 3 instrument code.

```
import ("music.lib");

vol = hslider ("amp", 0.1, 0, 1, 0.001)  : lp(10);
freq = hslider ("fund [unit:Hz]",440,110,880,1) : lp(10);
cf = hslider ("cf [unit:Hz]",1000,600,6000,1) : lp(10);
beta = hslider ("beta", 1,0,1,0.001) : lp(10);

lp (freq) = * (1 - a1) : + ~ * (a1) with
   { C = 2. - cos (2.*PI*freq/SR);
      a1 = C - sqrt (C*C - 1.); };
rms = fabs : lp(10);
balance (sig, comp) =  sig*rms (comp) / (rms (sig) +0.000001);
fbam (beta)  = ((+(1)) * (_)~* (beta));
k = int (cf/freq);
g = cf/freq - k;
mod = (1-g)*osci (k*freq)+g*osci ((k+1)*freq);
process = vol* (osci (freq) <: mod*fbam (beta),_ : balance);
```

A fourth variation on FBAM is obtined by inserting a non-linear wave-shaping function into the feedback path. The general form of this variant is given by

$$y(t) = \cos(\omega_0 t)[1 + f(\beta y(t-1))] \qquad (5.52)$$

where $f()$ is a non-linear function of some description. A good choice for this wave shaper is a cosine or a sine. In this case, we have

$$y(t) = \cos(\omega_0 t)[1 + \sin(\beta y(t-1))] \qquad (5.53)$$

We should note that this is very similar to some of the expressions we have encountered before in the recasting of FM/PM equations in the earlier sections of this chapter. Consider, for instance, the PM equation in which the modulator is the feedback signal $y(t-1)$:

$$y(t) = \cos(\omega t + \beta y(t-1))$$
$$= \cos(\omega t)\sin(\beta y(t-1)) - \sin(\omega t)\sin(\beta y(t-1)) \tag{5.54}$$

From this, we can see that the sine wave-shaped FBAM is equivalent to the first term of the expansion (plus the cosine input), therefore it partially implements this type of PM[8]. The result is a spectrum with no even harmonics, as depicted in Fig. 5.19.

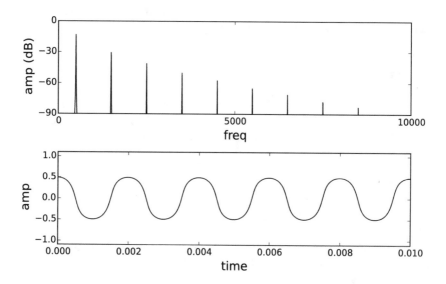

Fig. 5.19 Waveform and spectrum of FBAM variation 4.

5.4.2 Periodic time-varying filters

The fact that FBAM can be cast as a processing operation on a sinusoid (a PLTV filter) allows us to reimagine it as an adaptive method in a similar vein to the ones explored in Sects. 5.1 and 5.3. This can be implemented simply by replacing the sinusoid by an arbitrary input as in

---

[8] This is actually Feedback FM (or PM), which will be explored in section 5.4.3.

$$y(t) = x(t)[1 + \beta y(t-1)] \tag{5.55}$$

where $x(t)$ is any audio signal. In this case, we need to determine the amount of modulation employed according to the input signal. In some cases, we might be able to push β beyond the usual values used in basic FBAM. For noise-like sources, the filter is not PLTV anymore and the theory developed in Sect. 5.4.1 does not apply.

To realise an instrument based on this, we can use Faust to implement Eq. 5.55 in a Csound instrument, to which we can feed an input as in the other examples of adaptive methods. An example of this design is shown in Listing 5.18.

Listing 5.18 Adaptive FBAM instrument example.

```
gifbam faustcompile {{
  beta = hslider("beta", 0, 0, 10, 0.001);
  fbam(b) = ((+(1))*(_)~*(b)));
  process = fbam(beta);
}}, "-vec -lv 1"

instr 1
  ain = diskin:a("flutec3.wav",p5/cpspch(9.00))
  ib,asig faustaudio gifbam,ain
  kb = linen(p6,4,p3,3)
  faustctl(ib,"beta",kb)
  asig = balance(asig,ain)
     out(asig*p4)
endin
schedule(1,0,7,0.8,440,3)
```

The spectrum obtained from adaptive FBAM to transform a flute tone is plotted in fig 5.20, alongside its input for comparison.

First-order PLTV

In all examples so far, we have relied on using feedback as the modulation source. We can modify FBAM to create a generalised coefficient-modulated IIR filter by using two different signals as input and modulator. In order to keep it as a PLTV system, the modulation needs to be periodic. In the case of sinusoidal input and modulator signals, the expression becomes

$$y(t) = \cos(\omega_c t) + \beta \cos(\omega_m t) y(t-1) \tag{5.56}$$

where we will have a spectrum containing the input frequency $\omega_c (= 2\pi f_c)$ complemented by upper and lower sidebands at $\omega_c \pm \omega_m$, producing a double-sided set of partials. As before, $c : m$ ratios composed of small integers will produce harmonic spectra, while ratios of larger numbers result in a set of inharmonic components.

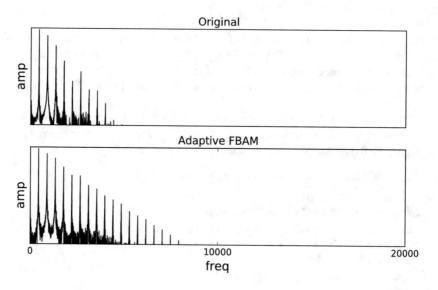

Fig. 5.20 Adaptive FBAM of a flute tone (original and transformed spectra).

The result is very similar to the spectra obtained from the double-sided generalised summation formula and PAF algorithm in Chap. 4, as illustrated in Fig. 5.21.

Equation 5.56 can be also used in adaptive applications by replacing the cosine input with an arbitrary signal. In this case, we can, as before, track the input pitch to control the modulation frequency according to a set $c : m$ ratio. These ideas are implemented in Listing 5.19, where Faust is again used to implement the filtering of an input signal that is pitch tracked by Csound. A flowchart for the Faust program is shown in Fig. 5.22.

Listing 5.19 Adaptive PLTV instrument example.

```
gifil faustcompile {{
import("music.lib");
beta = hslider("beta", 0, 0, 10, 0.001);
fm = hslider("fm", 80, 0, 4000, 0.001);
pltv(beta,fm) = _ + _ ~ *(osci(fm)*beta);
process = pltv(beta,fm);
}}, "-vec -lv 1"

instr 1
 irat = p7
 ain = diskin:a("flutec3.wav",p5/cpspch(9.00))
 ib,asig faustaudio gifil,ain
 kb = linen(p6,4,p3,3)
 kcps,kamp ptrack ain,512
```

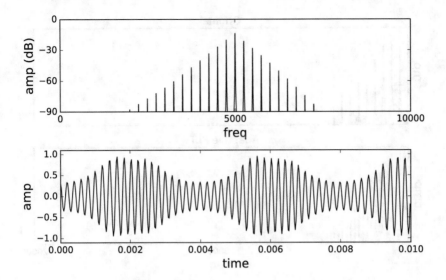

Fig. 5.21 Spectrum and waveform of a cosine-modulated, cosine-input, first-order PLTV filter, with $f_c = 5000$, $f_m = 250$ and $\beta = 0.9$.

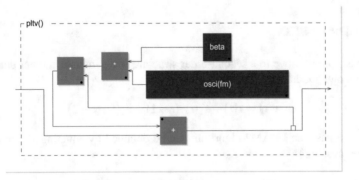

Fig. 5.22 First-order sinusoidal-modulation PLTV filter.

```
faustctl(ib,"beta",kb)
faustctl(ib,"fm",kcps/irat)
asig = balance(asig,ain)
    out(asig*p4)
endin
schedule(1,0,7,0.8,440,0.9,2)
```

This instrument has a similar behaviour to other adaptive designs seen earlier in this chapter. In Fig. 5.23, we can observe the effect of setting the $c : m$ ratio to 2, resulting in an octave transposition ($\beta = 0.9$).

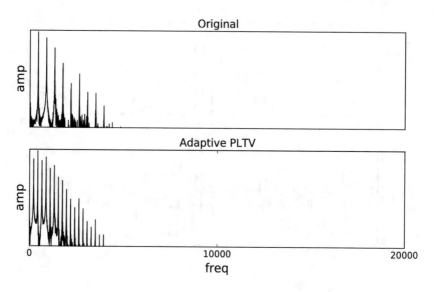

Fig. 5.23 Adaptive PLTV output spectrum for a processed flute tone compared with the original signal, with $c : m = 2$ and $\beta = 0.9$.

Second-order PLTV

It is possible to move to a second-order structure if we include another delay term with its own modulator [50]:

$$y(t) = x(t) + a_1(t)y(t-1) + a_2(t)y(t-2) \tag{5.57}$$

In this case, an FBAM2 algorithm can be obtained by equating $x(t) = a_1(t) = a_2(t) = \cos(\omega_0 t)$:

$$y(t) = \cos(\omega_0 t)[1 + y(t-1) + y(t-2)] \tag{5.58}$$

which behaves in a similar way to the first-order version, but produces a narrower pulse waveform and a richer spectrum (Fig. 5.24).

To allow a varying timbre, we can reintroduce the β parameter, but now affecting both feedback terms. While separate scaling could be applied to each delay, this does not provide enough extra timbral control to justify it. We can implement this design directly as a Faust instrument, shown in Listing 5.20. A flowchart for the PLTV filter is shown in Fig.5.25.

Listing 5.20 FBAM2 instrument.

```
import("music.lib");

vol = hslider("amp", 0.1, 0, 1, 0.001)   : lp(10);
```

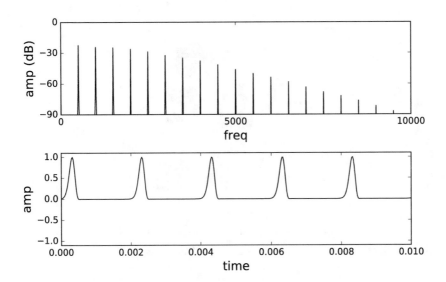

Fig. 5.24 Waveform and spectrum of FBAM2.

```
freq = hslider("freq [unit:Hz]",440,110,880,1) : lp(10);
beta = hslider("beta", 1,0,1,0.001) : lp(10);

lp(freq) = *(1 - a1) : + ~ *(a1) with
    { C = 2. - cos(2.*PI*freq/SR);
      a1 = C - sqrt(C*C - 1.); };
rms = fabs : lp(10);
balance(sig, comp) = sig*rms(comp)/(rms(sig)+0.000001);
fbam2(beta) = ((+(1))*(_) ~ (_ <: _ + _')*(beta));
process = vol*(osci(freq) <: fbam2(beta),_ : balance);
```

Similarly to the first-order case, it is possible to create a number of variants, including the use of a PLTV filter in a more general form, with the modulator and input signals decoupled. We can then proceed to look at existing types of second-order filters that can be put into a PLTV form. One case to consider is the basic resonator discussed in Sect. 3.4, which can be defined by

$$y(t) = x(t) - 2R\cos(\theta)y(t-1) - R^2 y(t-2) \qquad (5.59)$$

where the parameters θ and R, called the *pole angle* and *radius*, are determined by the filter centre frequency and bandwidth, respectively. In the PLTV case, the synthesis expression is

$$y(t) = x(t) - 2R\cos(\pi[\beta a(t) + \alpha])y(t-1) - R^2 y(t-2) \qquad (5.60)$$

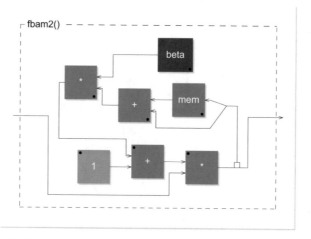

Fig. 5.25 Second-order PLTV filter used in the FBAM2 algorithm.

where $x(t) = a(t) = \cos(\omega_0 t)$ implements the modulation of the pole angle [50].

There are three parameters to work with: the pole radius R, the amount of modulation β and the angle offset α. A number of waveform shapes and spectra can be obtained with these parameters. For instance, with $R = 0.5$, $\beta = 1$ and $\alpha = 0$, we can produce a square wave. If we increase R, more harmonics will be generated, but aliasing can then become a problem.

Resonators also allow us to implement FM, although within a narrower range of parameters. If we set R close to or equal to 1 and limit β to values around 0.01, we can obtain α from a carrier frequency f_c in Hz (and the sampling rate f_s):

$$\alpha = \frac{2 f_c}{f_s} \tag{5.61}$$

However, this parameter setting can be unstable. Certain $c : m$ ratios, where m is the input sinusoid frequency, are impossible. The cases $c \leq m$ are problematic and some ratios of small numbers can also present difficulties (e.g. 3:2, 2:1). An interesting aspect is that the FM spectrum is only present for a short duration, determined by the decaying exponential R^t (with $R < 1$). This allows us to have an inharmonic onset based on a certain $c : m$ ratio, leading, after a certain amount of time, into a periodic tone defined by the pole angle modulation synthesis. We can also use different signals for the input x(n) and the modulator a(n). In this case, we can apply the techniques explored earlier in this chapter in another variant of AdFM.

Going one step further, the filter modulation can be applied to its frequency and bandwidth (rather than to the pole angle and radius). In this case, the modulating signal has to be converted into the parameters R and θ. For a more stable behaviour, we can keep a fixed Q ratio, which leads to both R and θ being modulated. In this case, we employ the following filter expression [50]:

$$y(t) = x(t) + a_1(t)y(t-1) - a_2(t)y(t-2) \tag{5.62}$$

where

$$
\begin{aligned}
r(t) &= \exp\left(\frac{\pi}{f_s}\frac{f(t)}{Q}\right) \\
a_1(t) &= \frac{4r(t)^2}{1+r(t)^2}\cos(\frac{2\pi f(t)}{f_s}) \\
a_2(t) &= r(t)^2
\end{aligned}
\tag{5.63}
$$

The modulation signal $f(t)$ can be constructed from a given carrier f_c and a modulator frequency f_m:

$$f(t) = f_c + A\cos(2\pi f_m t) \tag{5.64}$$

Care needs to be taken with the filter Q and with the frequency deviation A to maintain stability and avoid aliasing. It is possible to set $A = I f_m$, to put it in terms of a modulation index I. Resonator FM is more stable than pole-angle modulation and can be used more generally for a variety of effects. A simple example is given by feeding the filter with a sinusoid at f_c Hz, the result of which is shown in Fig. 5.26.

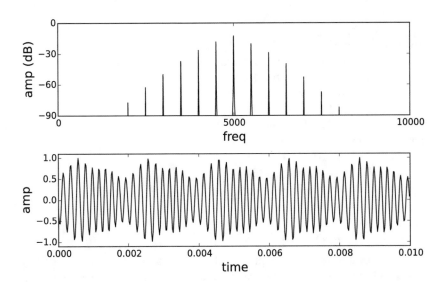

Fig. 5.26 Resonator FM (sinusoidal input) waveform and spectrum, with $f_c = 5000$ Hz, $f_m = 500$ Hz, $I = 2$ and $Q = 2.5$.

More generally, we should be able to use this PLTV filter to transform arbitrary audio signals. Again, with the insights of adaptive FM, we can track the input pitch and use it to control the modulator. The instrument in Listing 5.21 demonstrates the idea.

Listing 5.21 Resonator FM instrument.

```
gifl faustcompile {{
    import("music.lib");

    fc = hslider("fc",0,0,20000,1);
    fm = hslider("fm",0,0,20000,1);
    Q = hslider("Q",2.5,0.1,20000,1);
    I = hslider("I",0,0,20000,1);
    f = fc + I*fm*osci(fm);
    r = exp(-PI*(f/Q)/SR);
    cost = cos(2*PI*f/SR);
    b1 = (4*r*r/(1+r*r))*cost;
    b2 = -r*r;
    process = _ + _ ~ (_ <: (*(b1) + b2*_')) ;
}}, "-vec -lv 1"

instr 1
 indx = p6
 irat = p7
 iQ = p8
 kndx = linen(p6,4,p3,3)
 ain = diskin2:a("flutec3.wav",p5/cpspch(9.00))
 kcps,kamp ptrack ain,512
 ib, asig faustaudio gifl,ain
 faustctl(ib,"fm",kcps/irat)
 faustctl(ib,"fc",kcps)
 faustctl(ib,"I",kndx)
 faustctl(ib,"Q",iQ)
 asig = balance(asig, ain)
 aenv = linen(p4,0.1,p3,0.1)
    out asig*aenv
endin
schedule(1,0,7,0dbfs/2,440,2.5,1,2)
```

As in a previous example, we implement the FM resonator in Faust and employ it in a Csound instrument, where we can drive it with an arbitrary monophonic signal. The flute tone used in previous examples is shown alongside its transformation in the spectra plotted in Fig. 5.27. In this case, an inharmonic spectrum is generated by using an irrational value for $c : m$ ($\sqrt{2}$).

The principle of filter FM can be applied to other structures (such as Butterworth filters, etc.), as it is based on the simple idea of parameter modulation. However, as

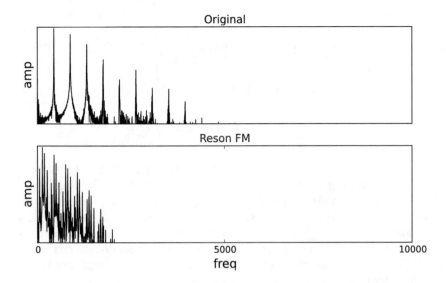

Fig. 5.27 Resonator FM applied to a flute tone, original and transformed spectra, with $c : m$ set to $\sqrt{2}$.

with a resonator, not all combinations of $c : m$, I and Q are possible, as the question of stability is brought to the fore. For this reason, care and experimentation are required to tune the parameters for specific requirements or desired results.

Higher-order filters

Higher-order PLTV filters can be built by allowing the feedback delay of the first-order PLTV form to be longer than a single sample [38, 49]. In the case of FBAM, we can make this delay variable:

$$y(t) = \cos(\omega_0 t)[1 + \beta y(n - d)] \tag{5.65}$$

where d is the delay length in samples.

From the point of view of a PLTV system, this defines a coefficient-modulated comb filter (with a sinusoidal input). The value of d has an effect on the output spectrum. A variety of waveforms can be produced with various delays. However, the spectrum will be invariant if the delay time (T_d) is locked in a fixed ratio to the modulation frequency (f_0 in the case of FBAM). As a side-effect, this allows us to keep the basic FBAM spectrum f_0-invariant if we increase the delay as the fundamental frequency decreases.

A case worth exploring is given by $T_d : f_0 = 1$, where $d = f_s/f_0 = \frac{2\pi}{\omega_0}$. This simplifies the expansion of the FBAM equation:

$$y(t) = \cos(\omega_0 t) \left[1 + \beta y \left(n - \frac{2\pi}{\omega_0}\right)\right]$$

$$= \sum k = 0^\infty \beta^k \prod_{\tau=0}^{k} \cos(\omega_0 t - 2\pi\tau) \tag{5.66}$$

$$= \sum_{k=1}^{\infty} \beta k - 1 \cos(\omega_0 t)^k = \frac{\cos(\omega_0 t)}{1 - \beta \cos(\omega_0 t)}$$

with $0 \le \beta < 1$. The fact that FBAM can have a closed form such as the one in Eq. 5.66 demonstrates its connections to the techniques developed in Chap. 4.

For its general implementation, this technique requires some form of delay-line interpolation. For delays that are longer than a few samples, a linear interpolation method might be sufficient, but for shorter delays, we will require a more precise, higher-order interpolator function. The aliasing characteristics of this form of FBAM are not too dissimilar to the first-order case, but the system might become unstable for larger β values in the specific cases of $d = \frac{f_s}{f_0}$ and $d = \frac{f_s}{2f_0}$.

5.4.3 Feedback FM

The principle of feedback modulation can also be applied to the frequency, or more precisely the phase, of a signal[9]. In this case, we have an algorithm that feeds the output back to modify the phase of a carrier oscillator, expressed by [102]

$$y(t) = \sin(\omega_0 t + \beta y(t-1)) \tag{5.67}$$

where as before $\omega_0 = 2\pi f_0$, for a fundamental frequency f_0.

It can be shown that the phase of this signal is described by

$$\phi(t) = \omega_0 t + \sin(\beta \phi(t-1)) \tag{5.68}$$

in which case the expansion of $\sin(\phi(t))$ becomes

$$2 \sum_{k=1}^{\infty} \frac{J(k\beta)}{k} \sin(k\omega_0 t) \tag{5.69}$$

This makes the feedback FM expression another very interesting closed form of a sine summation. Moreover, for $\beta = 1$, the Bessel-function-based coefficients can be approximated very handily as

$$2 \frac{J(k)}{k} \approx \frac{1}{k} \tag{5.70}$$

---

[9] We will use the term feedback FM which is more commonly associated with the technique

This renders the FM feedback spectrum very close to a sawtooth waveform, as we can observe in fig 5.28. By modifying β, as in feedback AM, we can obtain dynamic spectra. Over-modulating will make the system unstable, and will also cause objectionable aliasing as the formula is non band-limited. We also need to limit β at high fundamentals for this reason.

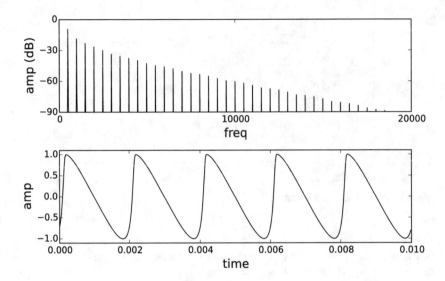

Fig. 5.28 Feedback PM waveform and spectrum.

A feedback FM instrument in Faust is shown in Listing 5.22. This implementation uses Eq. 5.68 for the phase feedback function $\phi(\omega)$, which distorts the linear phase term $\omega(t) = 2\pi[ft \bmod 1]$ (provided by the phase generator) (see Fig. 5.29). This form is equivalent to Eq. 5.67.

Listing 5.22 Feedback FM instrument.

```
import("music.lib");

vol = hslider("amp", 0.1, 0, 1, 0.001)   : lp(10);
freq = hslider("freq [unit:Hz]",440,110,880,1) : lp(10);
beta = hslider("beta", 1,0,1,0.001) : lp(10);

lp(freq) = *(1 - a1) : + ~ *(a1) with
   { C = 2. - cos(2.*PI*freq/SR);
     a1 = C - sqrt(C*C - 1.); };

mod1(a) = a - floor(a);
incr(freq) =  freq / float(SR);
```

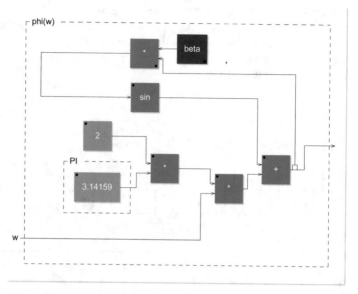

Fig. 5.29 Phase feedback modulation.

```
phasor(freq) =  incr(freq) : (+ : mod1) ~ _ ;
phi(w) = (_ : sin) + w  ~ *(beta);
fbfm(f) = sin(phi(2*PI*phasor(f)));
process = vol*fbfm(freq);
```

Delay-line feedback FM

Feedback FM can also be implemented through the delay-line method. In this case, recalling Eq. 5.14 in Sect. 5.2.2, we need to set the maximum deviation of the delay Δ_d from the parameter β as

$$\Delta_d = \frac{\beta}{\pi f_0} \tag{5.71}$$

With this setting, we can proceed to implement an equivalent version of the instrument in Listing 5.22 using a variable delay line. This is shown as a Csound instrument in Listing 5.23. A few remarks are due. The use of feedback requires a sample-by-sample operation, so we need to set the processing block size *ksmps* to 1. This is done by using `setksmps(1)` at the top of the instrument. The modulator feedback signal is inverted and offset (a minimum delay of two samples is needed for the cubic interpolation to be performed properly). Two envelopes, for timbre and amplitude, are employed.

Listing 5.23 Delay-line feedback FM instrument.

```
instr 1
 setksmps(1)
 aout init 0
 ib = p6/($M_PI*p5)
 as  = oscili(1,p5);
 kb = linen(ib,0.1,p3,0.4)
 adt = kb*(1-aout)/2 + 2/sr
 adp = delayr:a(1)
 aout = deltap3(adt)
    delayw(as)
 asig = linen(p4,0.01,p3,0.1)
       out(aout*asig)
endin
schedule(1,0,5,0dbfs/2,440,1)
```

Of course, the main reason for implementing the delay-line version of the algorithm is so that we use an arbitrary input signal witg it. We can replace the sinusoid
in the instrument in Listing 5.23 with a different input and the instrument should
work in a similar way. The example in Listing 5.24 employs a flute tone for this
purpose (Fig. 5.30 compares the input and output of this process).

Listing 5.24 Delay-line feedback FM instrument using a flute tone as input.

```
instr 1
 setksmps(1)
 aout init 0
 ib = p6/($M_PI*p5)
 as  = diskin:a("flutec3.wav",p5/cpspch(9.00));
 kb = linen(ib,0.5,p3,0.4)
 adt = kb*(1-aout)/2 + 2/kr
 adp = delayr:a(1)
 aout = deltap3(adt)
    delayw(as)
 aenv = linen(p4,0.01,p3,0.1)
       out(aout*aenv)
endin
schedule(1,0,5,0dbfs,440,1)
```

As we did in the case of self-modulation in Sect. 5.2.3, we can optionally track
the f_0 to provide a reference for the modulation width. This can then be used to set
the amount of delay variation according to β, in the case of monophonic signals.
However, in the absence of a pitch reference, this value can also be estimated or
experimented with. Unlike the case for AdFM, as the modulation is provided by
feedback, there is no $c : m$ ratio to be set. This may possibly allow a wider range of
input signals to be used.

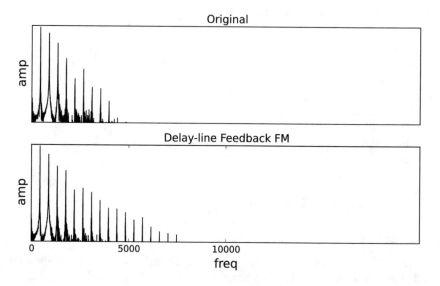

Fig. 5.30 The spectrum of delay-line feedback PM using a flute tone as input.

5.5 Conclusions

In this chapter, we explored a number of novel techniques that can be used to modify existing sounds, as well as to synthesise various types of spectra. We looked at the use of delay lines in various settings, employing modulation and feedback to create different types of instruments. In particular, we saw how adaptive methods can be employed to extract information, such as pitch, from an input signal in order to control the process.

Alongside a detailed look at the theory that supports these ideas, a variety of instrument examples were provided, both in Csound and in Faust, as well as some mixed code where the two languages were used in a complementary way. For cases where sample-by-sample signal processing is required (in the implementation of filters, for instance), we saw that Faust programs can be a good option for programming. Although Csound can also be used with block sizes of one sample, it is at times more efficient to take advantage of the native processing afforded by Faust.

Chapter 6
Granular Methods

Abstract In this chapter, we examine the technique of granular synthesis from first principles. We begin by looking at the definition of a grain, its fundamental characteristics and a basic instrument that can be used to synthesise it. Two basic types of grain generation are studied, synchronous and asynchronous, which can produce streams and clouds of particles. Example code implementing these techniques is presented. The chapter concludes with a look at grain generator opcodes in Csound.

Granular synthesis is the name given to a series of techniques based on the generation and manipulation of short segments of sound called *grains*. Such snippets are shaped by an envelope function that smoothly fades the grain in and out. A collection of a great number of such segments makes up the output of these granular methods. These methods allow the creation of varied types of sound textures and spectra, which can be flexibly and dynamically modified. In this chapter, we will introduce the technique from basic principles, examining what a grain is and the ways in which we can generate a collection of them.

6.1 Grains

The first step in studying granular synthesis is to define more precisely what we mean by a *grain*. Typically, we might describe an audio grain as a very short sound, which might in itself be featureless, a mere click or ping if played in isolation. When many of these are played over a certain duration, however, a number of sonic characteristics will emerge. An example is shown in Fig. 6.1, where we have what we could describe as a textbook grain, lasting for a small fraction of a second, containing a defined waveform (in this case a sine), and shaped by a smooth envelope.

Grains are defined by the following three characteristics, which have an influence on the overall sound when many grains are accumulated:

© Springer International Publishing AG 2017
V. Lazzarini, *Computer Music Instruments*,
https://doi.org/10.1007/978-3-319-63504-0_6

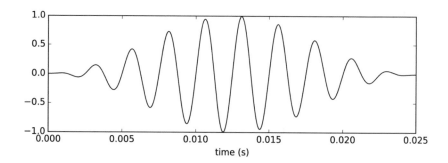

Fig. 6.1 An audio grain of 25 ms duration containing a sinusoidal waveform at 400 Hz.

1. **Envelope**: different types of amplitude shaping create a variety of spectral qualities. For example a short attack and decay will render a grain close to an impulse; a longer-lasting decay will reveal more of the grain waveform.
2. **Waveform**: the waveform is the raw content of a grain, which can be as simple as a sinusoid, or noise-like, or derived from an existing sound (e.g. a recording).
3. **Duration**: this parameter is associated with the grain envelope, although we tend to consider it separately. It defines whether a separate grain is closer to a click or an identifiable piece of audio. Granular synthesis tends to work with shorter durations, where each individual particle is not very recognisable, but when played alongside others allows complex spectra to emerge.

Envelope shapes can be symmetrical or asymmetrical, and can be defined by piecewise functions, with linear or non-linear segments, or by other window functions. The grain in Fig. 6.1 is made out of an inverted raised cosine (a von Hann window), whereas in Fig. 6.2 we see examples of other types of envelopes: asymmetric exponential, linear-piecewise (trapezoid), triangular and Gaussian. Each envelope shape can imply some extra parameters of its own, such as attack, decay, etc. In some cases, it is useful to call these *local* envelope parameters, to distinguish them from global ones that shape the overall sound composed of several grains.

The grain contents, its waveform, can also contribute significantly to the resulting sound. If the grain holds a sinusoid or some other simple wave shape, then an important parameter is its fundamental frequency. If the contents are an existing sound from a recording, then it is possible to replace this parameter by a transposition factor (a playback speed). For a single grain, the usual considerations with regard to aliasing will hold: that the signal should not contain components beyond $f_s/2$. However, when played as a stream, other considerations will arise, as, depending on the rate of grain generation and the shape of the envelope, aliasing distortion might emerge.

The lower limit on the grain duration is usually defined by a threshold between a grain and a waveform cycle. A very short grain will be very similar to a single period of a pulse wave. Durations can range from a few milliseconds to around half

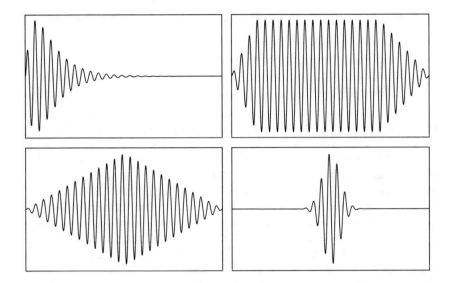

Fig. 6.2 Four different grain envelope shapes. Top row: exponential attack-decay; trapezoid. Bottom row: triangular; Gaussian.

a second, although at that end of the range it is debatable whether we are actually using grains or full segments of an audio signal. In this case, grains will retain enough of the quality of the source sound to be considered as, for instance, a short note with a given envelope.

It is straightforward to build an instrument that produces a single grain. All we need is an envelope and a signal generator. If we keep our waveform and our envelope function in a table, then all we need are two oscillators. The first one takes the reciprocal of the grain duration as the frequency and reads an amplitude-shaping function. The second one takes this signal as its amplitude and uses the grain waveform frequency, reading from a wave table (Listing 6.1).

Listing 6.1 Grain instrument.

```
instr 1
 aenv = poscil(p4,1/p3,p7)
 asig = poscil(aenv,p5,p6,p8)
   out(asig)
endin
```

Note the use of `poscil` instead `oscili`. These two opcodes are equivalent; however, the former is slightly more flexible in that it allows us to use function tables of any size (whereas the latter requires power-of-two lengths). With such opcodes, we can read wave tables containing recordings of different sounds (and not only single waveforms). The parameters for this instrument are, from p4: the amplitude, grain frequency, grain waveform, envelope waveform, and grain waveform phase (offset).

6.2 Grain Generation

When we consider the generation of grains, other parameters emerge. The first of these can be defined as the grain *rate* or frequency, which defines how often they are created. This is mostly applicable if we have a steady *stream* of grains, generated one after the other. Another way to look at this is to consider the grain density, which is an average number of grains over a certain period. This describes the process a little better if we are not thinking of streams but of *clouds* of them, statistically distributed in time. In this case, we do not necessarily know exactly when a grain will follow another, but we have an idea of their overall density (over, say, a second).

These two fundamentally different ways of generating grains are often known as *synchronous* and *asynchronous* granular synthesis. While the first can be used for a number of different harmonic and inharmonic spectral types, the second one is mostly used to create textures with distributed spectra. With these basic principles in mind, we can start experimenting with grain generation from first principles.

6.2.1 Grain streams

The idea is that we can employ the instrument in Listing 6.1 together with some scheduling code and generate a stream or a cloud of grains. Let's begin by considering the following parameters:

1. **Grain waveform**: sine.
2. **Grain envelope**: Gaussian function (see Fig. 6.2).
3. **Grain duration**: $\frac{1}{rate}$.
4. **Grain rate**: 300 grains/s.
5. **Grain waveform frequency**: 2000 Hz.

The complete program for this is shown in Listing 6.2. With it, we generate 4000 grains, and each grain starts $\frac{1}{300}$ sec after the previous one, creating a synchronous stream. The Gaussian window is created using GEN 20, which can calculate a variety of different shapes. Note also that we need to set ksmps to 1 in order to get sample-level accuracy for the event scheduling[1].

Listing 6.2 Grain stream generator

```
ksmps=1
```

---

[1] Csound also has a sample-accurate mode that can be used to allow us to set the ksmps size to higher values.

```
instr 1
 aenv = poscil(p4,1/p3,p7)
 asig = poscil(aenv,p5,p6,p8)
   out(asig)
endin

ifgs = ftgen(0,0,16384,20,6,1)
icnt init 0
ist init 0
irt = 300
igs = 1/irt
while icnt < 4000 do
   schedule(1,ist,igs,0dbfs/4,2000,-1,ifgs,0)
   ist += 1/irt
   icnt +=` 1
od
```

The resulting sound has a harmonic spectrum with a peak at the grain waveform frequency. The partials are spaced by the grain rate (which determines the fundamental frequency of the sound). The envelope shape is very narrow, which makes the signal have a large bandwidth. This is illustrated in Fig. 6.3, where the spectrum and waveform of this sound are shown. The reason for this is that with a short envelope, the grains start to resemble pulses and these exhibit a large number of partials.

Note that the grains in this sound do not overlap as the grain duration is the reciprocal of the grain rate. Now let's consider the result of having a longer grain duration, with some overlap. If we set the duration to twice the grain generation period, we can see that the result will be fewer components, always concentrated around the carrier. The grains almost join each other, because the result of the change in duration is a longer/wider envelope (Fig. 6.4).

Applying a different envelope to the grain also has an influence on the output waveform and spectrum. For example, if we replace the Gaussian window with a triangle shape, using the following line,

```
ifgs = ftgen(0,0,16384,20,3,1)
```

keeping the other parameters as before, we have a wider spectrum, still with a peak at the grain waveform frequency, but with a different overall shape, as is evident in Fig. 6.5.

Turning now to the contents of the grain, if we employ a waveform with more than one harmonic, we can generate a series of evenly-spaced peaks at multiples of the grain frequency. Each peak will be scaled by the relative weight of each waveform partial. For example, with a wave with three components weighted by the reciprocal of the harmonic number,

```
ifw = ftgen(0,0,16384,10,1,0.5,0.25)
```

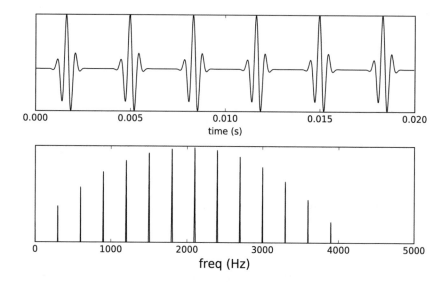

Fig. 6.3 Stream consisting of grains of $\frac{1}{300}$ sec generated at 300 Hz, containing a sine wave at 2 kHz, and using a Gaussian envelope.

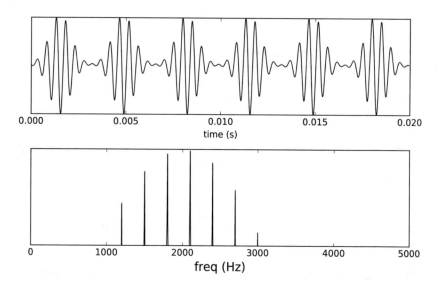

Fig. 6.4 Stream consisting of grains of $\frac{1}{150}$ sec generated at 300 Hz, containing a sine wave at 2 kHz, and using a Gaussian envelope.

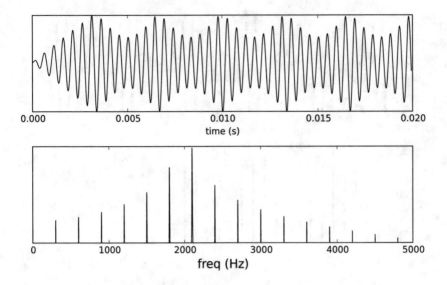

Fig. 6.5 Stream consisting of grains of $\frac{1}{150}$ sec generated at 300 Hz, containing a sine wave at 2 kHz, and using a triangle envelope.

we can have a spectrum with three peaks, at 2, 4 and 6 kHz as shown in Fig. 6.6, where we have used a von Hann (inverted, scaled cosine) envelope with a non-overlapping grain stream.

If we analyse what the code in Listing 6.2 is doing, we might be able to redesign the instrument to do it in a more versatile and efficient way. A non-overlapping grain stream consists of an enveloping waveform $w(t)$ that is ring-modulated by the grain waveform $g(t)$:

$$s(t) = w(t)g(t) \qquad (6.1)$$

Note that this formulation shows that although the grain wave looks at first sight like a sinusoid, this is not always the case. The grain consists of a signal that is segmented to fit the grain size, i.e. chopped off at the end. If the waveform period is not a multiple of the grain size, the output will have a spectrum that is very wide, created by the fact that the grain is incomplete. This is equivalent to applying a rectangular grain envelope to the waveform, this situation is shown in Fig. 6.7 where a sinusoid at 2000 Hz has its final cycle cut in half. The signal is not a sine wave any more, but a 300 Hz wave that is composed of 6.5 sine wave periods[2].

If there is a discontinuity at the end of the grain the stream does not have smooth joints, and then we will have aliasing. Thankfully, by using envelope waveforms that go to zero at the ends, this problem can be dealt with. With these ideas, we can re-create our grain stream generator based on the heterodyning operation of Eq. 6.1.

---

[2] Note the similarities between this and the hard-sync technique discussed in Chap. 3.

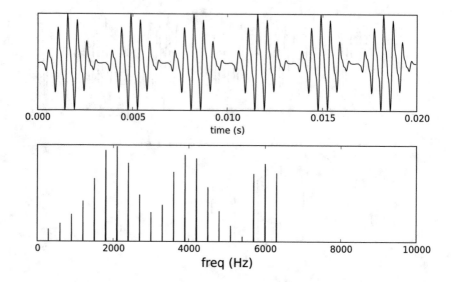

Fig. 6.6 Stream consisting of grains of $\frac{1}{300}$ secs generated at 300 Hz, containing a three-component wave at 2 kHz, and using a von Hann envelope.

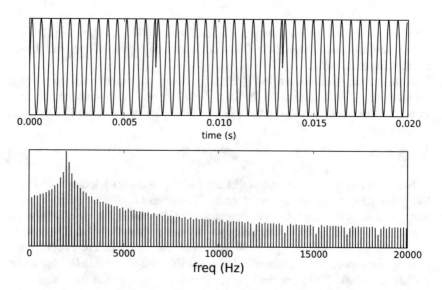

Fig. 6.7 Stream consisting of grains of $\frac{1}{300}$ sec generated at 300 Hz using a rectangular envelope.

To do this, we need to synchronise the phases of the two signals so that they start together, using a common phase source. Then we use two table readers to create the stream, one for each waveform, and multiply them together. The grain waveform frequency will be a multiple of the grain rate (not necessarily an integral multiple).

Listing 6.3 Synchronised grain stream generator

```
instr 1
  irt = p6/p5
  a1   = phasor(p5,p9)
  aenv = tablei:a(a1,p8,1,0,1)
  asig = tablei:a(a1*irt,p7,1,0,1)
    out(p4*asig*aenv)
endin

ifg = ftgen(0,0,16384,20,6,1)
schedule(1,0,60,0dbfs/2,300,2000,-1,ifg,0)
```

The parameters in this case are, from p4, the grain rate, grain waveform frequency, grain waveform, envelope shape, and phase offset. The last parameter is used if we want to create overlapping grains, so we can shift one stream against the other. For instance, to create a half-size overlap we schedule two instruments, with half the original grain rate, and offset the phase by 0.5. This instrument can generate the signals in figs. 6.3–6.6 more compactly and efficiently.

From these examples of synchronous granular synthesis, we can see that this basic form of the technique is capable of generating harmonic spectra of various kinds, with peaks determined by the grain waveform components. As with other techniques discussed in this book, we can vary some of its parameters to obtain dynamic spectra.

6.2.2 Grain clouds

Now we can consider the case where we grains are generated as a cloud rather than a stream. We can start with the following recipe:

1. **Grain waveform**: sine.
2. **Grain envelope**: Gaussian function.
3. **Grain duration**: 0.025 sec.
4. **Grain rate**: randomly distributed over 300 - 600 Hz.
5. **Grain waveform frequency**: randomly distributed over 1000 – 2000Hz.

In Listing 6.4, we can see the code for a sequence of grains that is generated according to this recipe.

Listing 6.4 Grain cloud generator

```
ksmps=1
```

```
instr 1
 aenv = poscil(p4,1/p3,p7)
 asig = poscil(aenv,p5,p6,p8)
   out(asig)
endin

icnt init 0
ist init 0
igs = 0.025
while icnt < 2000 do
    ifi = 300 + rnd(300)
    igf = 1000 + rnd(2000)
    schedule(1,ist,igs,0dbfs/4,igf,-1,if3,0)
    ist += 1/ifi
    icnt += 1
od
```

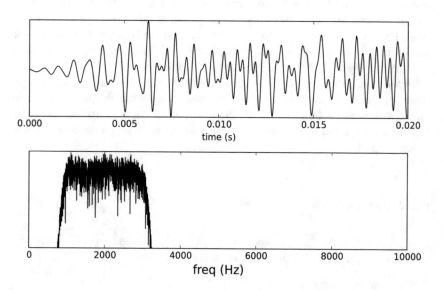

Fig. 6.8 Cloud of 0.025-sec grains of secs generated randomly with a density between 300 and 600 times per second, containing sine waves of various frequencies between 1000 and 3000 Hz, and using a Gaussian envelope.

The result is a distributed noisy spectrum around the waveform frequency range, as shown in Fig. 6.8. The quality of the sonic effect is not, however, captured greatly the static representation of the spectrum. Instead, the result has an interesting particle-like texture characteristic that is not simply the sound of band-limited noise. The spectrogram plot in Fig. 6.9 reveals more of the granular structure of this sound.

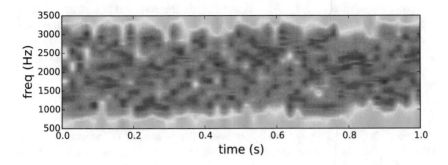

Fig. 6.9 Spectrogram of the sound generated by Listing 6.4 and shown as a waveform and spectrum in Fig. 6.8

Taking this simple example as a starting point, we can create different types of cloud textures by varying the grain size, which is now independent of the grain rate, the start time of each grain (instead of defining a grain rate); the range of grain frequencies, and the envelopes. All of these parameters can have different time evolutions to realise complex dynamically changing sounds. An example of this is shown by the sequence generated using the code in Listing 6.5, where we start with sparse grains and build up to a density of 1000 grains/sec.

Listing 6.5 Evolving grain cloud.

```
icnt init 0
ist init 0
igs = 0.04
iv init 1
iv2 init 1000
while icnt < 8000 do
    ifi = iv + rnd(iv)
    igf = iv2 + gauss(iv2/2)
    schedule(1,ist,igs,0dbfs/16,igf,-1,igf)
    ist += 1/ifi
    iv = iv > 1000 ? 1000 : iv + 0.5
    iv2 = iv2 > 5000 ? 5000 : iv2 + 10
    icnt += 1
od
```

6.2.3 Sampled-sound sources

As mentioned before, the sonic content of grains can be sourced from existing recordings (*sampled* sound). We can use exactly the same grain instrument (Listing 6.1) to read from a table that now contains a complete sound (rather than only a single waveform cycle). This can be created with GEN 1, which can read from a sound file and fill the table with the corresponding samples samples, for example,

```
ifw = ftgen(0,0,0,1,"fox.wav",0,0,1)
```

where the file fox.wav will be loaded onto the table.

For this application, the oscillator frequency needs to be adjusted to become a transposition factor. This is done by observing that to read all of the samples in this table from the beginning to the end we need to have a phase increment of 1 (see Sect. 3.2.1), which is, according to Eq. 3.3, equivalent to

$$\frac{N}{f_s} \tag{6.2}$$

where N is as the table size in samples. Therefore, the transposition factor t is this quantity times the required playback speed p:

$$t = p\frac{N}{f_s} \tag{6.3}$$

In the case where a complete sound is used as a source, we can play with the recognisability of the material by using different grain sizes. Longer grains will reveal the original sound, but short grains with enough density will also do the same, if we offset them to read consecutive positions in the file. So here's a recipe for a granulated resynthesis of an existing sound with some dynamic evolution:

1. **Grain waveform**: taken from different slices of a recorded sound, progressing from the beginning of the sound to the end as grains are generated.
2. **Grain envelope**: von Hann window.
3. **Grain duration**: varying from 0.5 to 0.05 secs.
4. **Grain rate**: random, but increasing in density from 1 to 1000 grains/sec.
5. **Grain waveform frequency**: randomly distributed around the original pitch.

This design starts with sparse grains of long duration and moves to short grains with a high density. We will hear snippets of the start of the sound, and then these will be taken over by a wide-bandwidth grain cloud, finishing with a granulated playback of the end of the sound. This becomes smoother as the density increases. The full code in Csound is presented in Listing 6.6, using a recording of the text 'the quick brown fox jumps over the lazy dog'. The word 'quick' gets repeated to start with, before the cloud dominates. We are then able to recognise the final part of the sentence with the growth in grain density.

Listing 6.6 Grain cloud obtained from a recording of 'the quick brown fox jumps over the lazy dog'.

```
ksmps=1
instr 1
 aenv = poscil(p4,1/p3,p7)
 asig = poscil(aenv,p5,p6,p8)
   out(asig)
endin

igf = ftgen(0,0,16384,20,2,1)
ifw = ftgen(0,0,0,1,"fox.wav",0,0,1)
icnt init 0
ist init 0
igs init 0.5
iv init 1
while icnt < 8000 do
   ifi = iv + rnd(iv)
   igf = 1 + gauss(0.15)
   schedule(1,ist,igs,0dbfs/2,igf*sr/ftlen(ifw),
              ifw,igf,icnt/8000)
   ist += 1/ifi
   iv = iv > 1000 ? 1000 : iv + 0.2
   igs = igs < 0.05 ? 0.05 : igs - 0.001
   icnt += 1
od
```

Note that the grain waveform phase, its reading position, moves linearly from the beginning of the table to the end, as it is indexed by the grain count `icnt`. We can introduce some randomness into this parameter, which will jumble up the grains and create some sort of stuttering effect. For instance, this code modification adds a random spread of 10% duration around the reading position:

```
ipos = icnt/8000 + gauss(0.1)
schedule(1,ist,igs,0dbfs/4,igf*sr/ftlen(ifw),
              ifw,ifl,ipos)
```

A huge variety of sound textures and effects can be created using sampled-sound sources. This technique offers great scope for the development of ideas for composition and sound design. To complement the discussion, we can make the grain cloud generator more compact and efficient by taking advantage of recursive scheduling in instruments. The example in Listing 6.7 builds on the previous ideas but implements them by making each grain instantiate the next one:

Listing 6.7 Recursive grain generation.

```
instr 1
 ibas = sr/ftlen(p6)
 iftl = 1/ibas
```

```
ifr =   p5*ibas
ips =   p8+gauss(0.25)
aenv =  poscil(p4,1/p3,p7)
asig =  poscil(aenv,ifr,p6,ips)
idens = p9 + gauss(p9)
ist = 1/idens
ist = ist < 1/kr ? 1/kr : ist
ips = p8 < 1 ? p8 : 1 - p8
schedule(1,ist,p3,p4,1+gauss(0.1),
         p6,p7,ips+ist/iftl,p9)
   out(asig)
endin

ifg = ftgen(0,0,16384,20,2,1)
ifw = ftgen(0,0,0,1,"fox.wav",0,0,1)

idens = 1000
igs = 0.05
schedule(1,0,igs,0dbfs/4,1,ifw,ifg,0,idens)
```

In this case, the frequency, read position and density are given a certain random spread. Each grain schedules the next at 1/dens secs later, and the grain position is also dependent on the current density and on the length of the original sound (ift1). We also keep the position modulo 1 to avoid the oscillator phase growing to stratospheric values. Notice that the maximum density is limited by the control period (1/kr, or ksmps × sr), as each grain has to be at least that distance from any other grain in time. This instrument can be used as a basis for further asynchronous granular synthesis designs.

6.2.4 Using Python to generate grain data

In the previous examples, we have generated all grain data using Csound code, which can be convenient for certain applications. Alternatively, we can employ the scripting possibilities of Python to perform this task. In this case, we modify the instrument slightly so that references to function tables are built into it. This will save us having to query Csound about function table sizes from Python. This reduces the number of instrument parameters to 6 (Listing 6.8).

Listing 6.8 Modified grain instrument.
```
instr 1
 ibas = sr/ftlen(gifw)
 aenv = poscil(p4,1/p3,gifg)
 asig = poscil(aenv,p5*ibas,gifw,p6)
   out(asig)
```

```
endin
```

To send the grain data to Csound we can use its *score* format, which is a list of statements, each starting with the letter 'i', and followed by the instrument parameters (p1, p2, p3, ...) separated by spaces. For instance,

```
i 1 0 1 16000  1 0.001
```

schedules an event for instrument 1 in Listing 6.8, starting at time 0, for 1 second, with 16000,1, and 0.001 for p4, p5, and p6, respectively.

This format can be used to construct a score containing all the grain data for a performance as a Python string. The Csound API function `readScore()` is used to load the score. We can then perform it to generate the audio. A full example program is shown in Listing 6.9

Listing 6.9 Modified grain instrument.

```python
import random as rnd
import ctcsound as csound

code = '''
ksmps=1
gifg = ftgen(0,0,16384,20,6,1)
gifw = ftgen(0,0,0,1,"fox.wav",0,0,1)
instr 1
 ibas = sr/ftlen(gifw)
 aenv = poscil(p4,1/p3,gifg)
 asig = poscil(aenv,p5*ibas,gifw,p6)
   out(asig)
endin
'''

def schedule(p2,p3,p4,p5,p6):
   return "i1 %f %f %f %f %f \n" % (p2,p3,p4,p5,p6)

cs = csound.Csound()
cs.setOption('-odac')
a = cs.get0dBFS()
sc = ""
st = 0.0
gs = 0.5
v = 1.0
tg = 16000
for i in range(0,tg):
   f = v + rnd.random()*v
   gf = 1 + rnd.gauss(0,0.04)
   pos = 2*i/tg + rnd.gauss(0,0.025)
   print(pos)
```

```
    sc += schedule(st,gs,a/4,gf,pos)
    st += 1/f
    if v < 1000: v += 0.1
    if gs > 0.05: gs -= 0.001

cs.readScore(sc)
cs.compileOrc(code)
cs.start()
cs.perform()
```

Python has a number of facilities for random number generation that can be quite useful for asynchronous granular synthesis. It can provide a flexible environment for creating complex evolving textures.

6.3 Grain Generators in Csound

While we have concentrated on exploring grain generation from first principles and designed instruments and sequencing to achieve this, it is also possible to employ granular synthesis opcodes for this purpose. These are self-contained unit generators that will produce streams or clouds (or both) according to a set of input parameters. In Csound there is a whole suite of such opcodes, from which we will select three for closer inspection: `syncgrain`, `fof`, and `partikkel`.

6.3.1 Syncgrain

This opcode, as implied by its name, is designed to generate streams of grains in a synchronous manner, although if random functions are used to control its parameters, it might be used for the synthesis of cloud-like textures. It works by creating separate grains sequentially, in a conceptually-similar manner to the code in Listing 6.2, based on the following principles:

1. Grains are separated by a given generation period, which is the inverse of the grain generation frequency.
2. At the start of each period, grains are created starting from a given position in a source table.
3. Grains have a set duration, playback rate and envelope shape.
4. Grains can overlap depending on the size and generation frequency.
5. The starting read position for each grain is set by a separate parameter.

From this recipe, `syncgrain` is defined by

```
asig = syncgrain(ka,kf,kp,ks,kt,ifw,ifg,iol)
```

where

ka – amplitude.
kf – grain generation frequency.
kp – grain waveform pitch (transposition ratio).
ks – grain size.
kt – read position increment, in whole grains.
ifw – source function table.
ifg – envelope function table.
iol – maximum number of overlapped grains.

Most of the above parameters are intuitively related to our previous discussion of granular synthesis in Sect. 6.2. The read position increment perhaps requires some further explanation. This determines where the next grain is going to be read from in the source function table and is defined in terms of the current grain duration. An increment of 1 will skip ahead one full grain size, so the read pointer will advance by a whole grain at a time. Fractional values move the pointer forward by less than a grain and values greater than 1 cause it to jump by more than one full grain.

This affects the timescale of playback: if there is a single stream, then with an increment of 1 we will proceed at the original speed, a value greater than 1 will stretch the duration, and a value less than 1 will compress it. An increment of 0 will freeze the playback at the current position, and a negative value will move the grain starting point backwards. For overlapping grains, which will be the case if $ks >$ $1/kf$, the increment needs to be divided by the overlap amount ($kf \times ks$). This is also the case if there are are gaps in the stream ($ks < 1/kf$).

As usual, the grain transposition (playback rate) parameter can also be negative for backwards playback. So the four parameters, frequency, grain size, pitch, and position increment can be combined together to create a variety of granular effects. In Listing 6.10, we see an example of the usage of this technique, where the same source used in earlier examples is manipulated.

Listing 6.10 Granular manipulation with syngrain

```
gifw = ftgen(0,0,0,1,"fox.wav",0,0,1)
gifg = ftgen(0,0,16384,20,2,1)
instr 1
  kf  = line(2,p3,100)
  ks  = line(0.5,p3,0.005)
  kol = ks*kf
  kt  = line(0,p3,0.5)
  kp  = 1 + randh(0.1, 1/ks)
  a1  = syncgrain(p4,kf,kp,ks,kt/kol,gifw,gifg,14)
     out(a1)
endin
schedule(1,0,60,0dbfs/2)
```

In this example, we develop a granular texture whose grains shrink from 0.5 to 0.005 secs, and the generation frequency grows from 2 to 100 grains per second. We also speed up while going through the table from a static position to a time stretch factor of 0.5. The grain pitch fluctuates between 0.9 and 1.1, varying at a frequency that is the reciprocal of the grain size. The texture builds to reveal the text of the speech at the end of the synthesis.

This opcode is capable of creating various types of synchronous granular synthesis sounds. Alongside it, Csound also offers two variants: `syncloop`, which allows one to create loops of function table data, and `diskgrain`, which creates streams directly from a sound file (rather than a function table).

6.3.2 FOF

FOF, Fonction d'Onde Formantique (formant wave function) [90], is a synthesis method originally designed for voice synthesis that shares some similarities with synchronous granular synthesis. The principle on which FOF is based is that, as we have seen in Sect. 6.2.1, a stream of sinusoidal grains has a harmonic spectrum with a peak centred on a given frequency (the grain waveform frequency). To synthesise a vowel, four or five of these streams are mixed together, each one defining one formant region.

The FOF *granule* (as it is called) is made up of an envelope consisting of a short attack and an exponential decay, with a final tail-off to 0 (Fig. 6.10). The exact shape of the envelope determines the spectral curve around the centre frequency. The rise time defines the width of a −40 dB region of the formant, whereas the decay sets its −6 dB bandwidth. A long, sustained, decay creates a narrow region, whereas a short, fast decay is correlated with a wideband spectrum. The waveform and spectrum in Fig. 6.11 illustrate a typical example of a FOF stream.

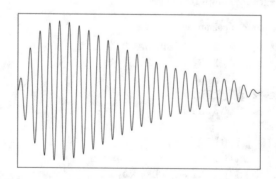

Fig. 6.10 FOF grain.

The Csound FOF generator is defined by

```
asig = fof(xa,xfo,xff,koc,kbw,kat,ks,kdc,iol,if1,if2,idr)
```

where

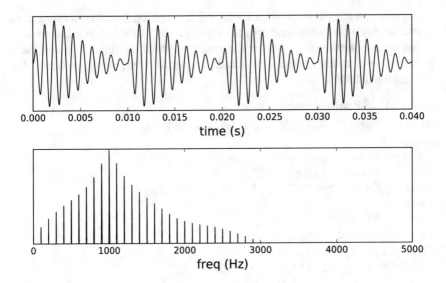

Fig. 6.11 FOF stream.

`xa` – amplitude.
`xfo` – fundamental frequency.
`xff` – formant frequency.
`koc` – octave transposition.
`kbw` – bandwidth.
`kat` – grain envelope attack time.
`ks` – grain size.
`kdc` – grain envelope decay.
`if1` – grain waveform function table.
`if2` – envelope attack shape.
`iol` – maximum number of overlapping grains.
`idr` – total performance duration.

The parameters match the FOF description given above, where the grain stream is produced at a fundamental frequency `xfo`, each grain containing a waveform with a frequency `xff` derived from the table `if1`. The grain size is `ks`, with an attack `kat`, and an exponential decay determined by the bandwdith `kbw`. Long decays are equivalent to small bandwidths and short decays to large ones. Each grain is tailed off by a short decay set by `kdc`. The attack shape is derived from the function table `if2`. The octave transposition parameter `koc` controls the fading out of consecutive grains, which makes the sound move smoothly an octave down for each integral value greater than 0 (e.g. `koct` = 1 produces a 1-octave downshift). As this places gaps between grains, it can be used to create a granulation effect without a pitch glide.

An example of FOF is shown in Listing 6.11, where a granular synthesis texture is created (with some similarities to the syncgrain example). Here, sinusoidal grains are accumulated over time until a continuous sound sliding upwards in pitch is created, with a resonance region around the formant frequency (randomly distributed around 500 Hz).

Listing 6.11 FOF granular synthesis example.

```
gif = ftgen(0,0,16384,5,0.001,16384,1)
instr 1
 kfo = line(1,p3,200)
 ks  = line(0.5,p3,0.005)
 kff = 500 + randh(50, 1/ks)
 koc = 0
 kat = 0.005
 kdc = ks*0.1
 kq = line(100,p3,10)
 a1 = fof(p4,kfo,kff,koc,kff/kq,
                  kat,ks,kdc,
                  100,-1,gif,p3)
     out(a1)
endin
schedule(1,0,60,0dbfs/4)
```

As mentioned earlier, FOF was designed originally to simulate vocal (vowel) sounds. In this application, we employ one fof generator for each formant region, mixing them to construct the full vowel. While our focus in this chapter has been to discuss the granular qualities of this method, it is important to note its other uses. In particular, FOF provides great scope for creating transitions between recognisable voice-like sounds and abstract granular textures. An example of these types of applications is provided in Appendix B (Sect. B.5).

In addition to fof, there are also two variations of this opcode that are designed to take advantage of its synchronous granular synthesis characteristics: fof2 and fog. These provide a means of dynamically selecting different read positions in the source table (as in syncgrain), allowing more flexibility in applying the FOF method to sampled-sound material.

6.3.3 Partikkel

The partikkel opcode, inspired by [87], is a unified implementation of all methods of time-domain granular processing [12], also known as *particle* synthesis. This is a very complex generator, with up to eight outputs and forty input parameters. The opcode exposes a huge number of facilities to manipulate grains. Many of its inputs are function tables that hold parameter changes acting on a per-grain basis.

It can employ single-cycle or sampled-sound material as sources for grains, which can be created as a mix of up to four wave tables.

A fifth source in `partikkel` is provided by *trainlet* synthesis, which generates an impulse train. These trainlets feature control of a dynamically-variable number of harmonics and balance between high and low frequencies, as well as of the fundamental frequency of the signal. Sources can be masked (muted) individually and on a per-grain basis. Different mixes can be sent to each one of the eight outputs.

Partikkel allows the creation of synchronous streams, as well as asynchronous clouds with random distribution of grain parameters. Source tables can be read from any start position, which can be modulated at audio rate, providing a sort of non-linear distortion effect. The envelopes of each grain can be defined precisely through a set of shape parameters. A full account of the various types of granular synthesis afforded by `partikkel` is given in [59, pp. 337–384].

6.4 Conclusions

Granular synthesis and processing techniques provide huge scope for the creation of interesting sounds and textures. They allow non-traditional ways of manipulating sound sources, which can be quite flexible and intuitive. This chapter has attempted to introduce the various concepts related to these techniques from first principles. We looked at examples that were built from a very simple instrument specification, which, given the correct sequencing of controls, can generate various types of sound, from pitched, resonant spectra to cloud-like grain bursts.

The range of possibilities is vast and this chapter has only touched on the foundations of these granular methods. However, grain manipulation is very open to experimentation. The programming examples provided here are a good starting point for exploration and the reader is encouraged to use them as a basis for the development of increasingly complex and sophisticated synthesis routines.

Chapter 7
Frequency-Domain Techniques

Abstract This chapter explores in detail the topic of spectral analysis and synthesis. It begins by looking at the concept of the Fourier transform and series and examines how we can derive the spectrum of a signal from its time-domain waveform. We then explore the discrete-time discrete-frequency version of the Fourier transform, the DFT, its formula and its implementation. This allows us to look in more detail into the characteristics of the spectral representation. Following this, we study the radix-2 fast Fourier transform algorithm, which is the most commonly used implementation of the DFT. The application of spectral processing to convolution and reverb completes the section. The second half of the chapter is dedicated to time-varying spectral analysis, where we introduce the short-time Fourier transform and its applications in spectral processing. We look at instantaneous frequency analysis and the phase vocoder, completing the study with a number of application examples. The chapter concludes with some notes on real-time spectral processing.

Frequency-domain techniques are based on the direct manipulation of a spectral representation of an audio signal. In this case, we are working with data that is primarily structured in terms of frequency, rather than time, unlike the case of the techniques explored in previous chapters. However, even though the focus is on the spectrum, we also need to consider the time dimension, as we take into account the dynamic evolution of a sound. Together, these two domains are manipulated through *time-frequency* methods, which we will examine later in this chapter.

To access the spectral data for manipulation, we need to employ an *analysis* step, transforming the waveform into its spectrum [6, 22, 68]. Conversely, when we want to obtain a signal for performance, it is necessary to employ a *synthesis* operation, converting spectra back into waveforms. We will examine the fundamental techniques for analysing and synthesising frequency-domain data and the main characteristics of this representation. We will follow this with an exploration of spectral data manipulation.

© Springer International Publishing AG 2017
V. Lazzarini, *Computer Music Instruments*,
https://doi.org/10.1007/978-3-319-63504-0_7

7.1 Frequency-Domain Analysis and Synthesis

In Chap. 1, we introduced informally the concept of breaking down waveforms into simple sinusoidal components of different amplitudes, frequencies and phases. This follows from a theory developed by Fourier [11, 26] that underpins many of the practical spectral signal-processing techniques. We will now look in a more formal way at some of the mathematical tools that have been derived from this theory and explore their particular characteristics.

The foundation for the theory is provided in the form of a continuous-time, continuous-frequency transform, from which a number of variations can be derived. Firstly, we have a continuous-time discrete-frequency version, a series, that can be used to define a periodic waveform. Then, in order to generalise the process to digital signals, we have a discrete-time, discrete-frequency version, which is the basis for many of the frequency-domain algorithms. In this section, we will try to explore each one of these with some application examples.

7.1.1 The Fourier transform

The fundamental mathematics for the type of spectral processing explored in this chapter is based on the Fourier transform (FT). Its formulation is continuous in time t, covering continuously all frequencies f from $-\infty$ to ∞ [98]. The FT can be expressed as follows [68]:

$$X(f) = \int_{-\infty}^{\infty} x(t)\left[\cos(2\pi ft) - j\sin(2\pi ft)\right]dt \tag{7.1}$$

where a spectrum $X(f)$ is obtained from a waveform $x(t)$, and f is given in Hz when t is defined in seconds. The FT works by multiplying the input signal by a sinusoid at frequency f, then performing an integral sum over this product[1], which makes it detect a sinusoidal component at that frequency if there is one. In this case, $X(f)$ denotes a pair of numbers, the spectral coefficients of f, indicating the amplitude of a cosine and/or a sine at that frequency. A sinusoid of any phase and amplitude at f can be uniquely described by these two quantities. Together, the cosine and sine coefficients make up a complex number.

The FT has an inverse transform, which starts with a spectrum $X(f)$ and then, through an integral sum, gives us a waveform at time t:

$$x(t) = \int_{-\infty}^{\infty} X(f)\left[\cos(2\pi ft) + j\sin(2\pi ft)\right]df \tag{7.2}$$

which can be considered as a synthesis expression, giving us a time-domain signal from its frequency-domain representation.

---

[1] See Appendix A.5 for more details.

The form of $X(f)$, as cosine and sine coefficients, can be converted into a more intuitive pair of functions, which gives us the amplitudes and phases of the components directly:

$$A = \sqrt{c^2 + s^2} \quad \text{and} \quad \phi = \arctan\left(\frac{s}{c}\right) \tag{7.3}$$

where A and ϕ are the amplitude and phase, respectively, corresponding to the cosine and sine coefficients[2] c and s. As a shorthand, we can also say

$$A(f) = |X(f)| \tag{7.4}$$

and

$$\phi(f) = \arg(X(f)) \tag{7.5}$$

for the amplitude (or magnitude) and phase spectrum, respectively. This is known as the *polar* form of the spectrum coefficients.

In the case of audio signals, which are real-valued, the magnitude spectrum is symmetric about 0 Hz and the phase spectrum is antisymmetric also about that frequency. This is helpful in that it allows us to derive the negative spectrum from its positive side and vice-versa, if we need to.

With periodic waveforms, as we have already seen, the frequencies f that make up the harmonic components are at multiples of a fundamental. This allows us to replace the integral by a summation in Eq. 7.2, giving rise to the Fourier series, defined for real signals as

$$x(t) = \frac{1}{\pi}\left[a_0 + 2\sum_{k=1}^{\infty} a_k \cos(2\pi kt) - b_k \sin(2\pi kt)\right] \tag{7.6}$$

where a_k and b_k are the sine and cosine amplitudes of each harmonic, which can be derived using Eq. 7.1 over one cycle:

$$
\begin{aligned}
a_k &= \frac{2}{\pi}\int_0^{2\pi} x(t)\cos(kt)\,dt \\
b_k &= -\frac{2}{\pi}\int_0^{2\pi} x(t)\sin(kt)\,dt
\end{aligned}
\tag{7.7}
$$

The spectrum is thus composed of all frequencies $f_k = kf_0$, from 0 Hz to ∞. In polar form, where A_k and ϕ_k are the amplitudes and phase offsets for each harmonic k, we have:

$$x(t) = \frac{2}{\pi}\left[\frac{A_0}{2} + \sum_{k=1}^{\infty} A_k \cos(2\pi kt + \phi_k)\right] \tag{7.8}$$

---

[2] Or, if we are using the complex form of $X(f)$, $c = \Re\{X(f)\}$ and $s = \Im\{X(f)\}$.

Let's look at an example: consider a bipolar sawtooth wave, defined in the interval $t = [0, 1]$ as

$$x(t) = (1 - 2t) \tag{7.9}$$

Using the definitions of the Fourier series coefficients in Eq. 7.7, we have

$$a_k = 2\pi \int_0^1 (1 - 2t) \cos(2\pi kt) dt = 0$$

$$b_k = -2\pi \int_0^1 (1 - 2t) \sin(2\pi kt) dt = \frac{-2\pi^2 k \cos(2\pi k) + \sin(2\pi k)}{2\pi^2 k^2} = -\frac{1}{k} \tag{7.10}$$

Substituting back in Eq. 7.6 gives a sawtooth wave defined by:

$$x(t) = \frac{2}{\pi} \sum_{k=0}^{\infty} \frac{1}{k} \sin(2\pi kt) \tag{7.11}$$

a result that can be verified by programming it using a large number of harmonics. This is demonstrated by the code in Listing 7.1, whose output is plotted in Fig. 7.1. Notice that because the signal is band limited, it overshoots its predicted absolute maximum value by a small amount at the transition points. This is known as the Gibbs effect [89]. It would not be present in a waveform with an infinite number of harmonics as determined by Eq. 7.11.

Listing 7.1 Sawtooth generation with the Fourier series.

```
import pylab as pl

saw = pl.zeros(10000)
t   = pl.arange(0,10000)
T = 5000
N = 1000
for k in range(1,N+1):
    saw += (1/k)*pl.sin(2*pl.pi*k*t/T)
saw *= 2/(pl.pi)

pl.figure(figsize=(8,3))
pl.plot(t/T, saw, 'k-', linewidth=2)
pl.tight_layout()
pl.show()
```

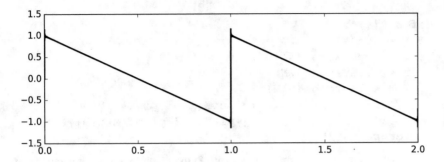

Fig. 7.1 Sawtooth wave produced by Eq. 7.11 using 1000 harmonics.

7.1.2 The discrete Fourier transform

So far the transforms and series we have been working with have been defined in continuous time. Now we need to transition to digital signals, which are time discrete. Furthermore, given that we want to represent the analysis data similarly in a sampled format, we require a tool that is also discrete in frequency. Thankfully, there is a version of the Fourier transform that is defined under such constraints. It is called the discrete Fourier transform [33, 34, 42, 76], and can be defined by the following expression:

$$X(k) = \frac{1}{N} \sum_{t=0}^{N-1} x(t) \left[\cos(2\pi k t / N) - j\sin(2\pi k t / N) \right] \qquad k = 0, 1, ..., N-1 \quad (7.12)$$

If we compare this with Eq. 7.1, it is possible to see that both time and frequency are not continuous, but limited to N discrete steps (N is called the DFT size). The frequency f is replaced by an index k, which refers to a set of frequency points. As we are sampling in time, our frequency content is bound by the Nyquist frequency $\frac{f_s}{2}$. As the spectrum ranges over negative and positive frequencies, the DFT will output N samples[3] covering frequencies from $-\frac{f_s}{2}$ to $\frac{f_s}{2}$.

The first $\frac{N}{2}$ of these are each linked to a frequency range from 0 to the Nyquist frequency. The $\frac{N}{2}$ coefficient corresponds to both $\frac{f_s}{2}$ and $-\frac{f_s}{2}$. The second half of the DFT output then consists of the negative frequencies from $-\frac{f_s}{2}$ to 0 Hz (excluding the latter value). We can program the DFT as a Python function to examine some of its characteristics (Listing 7.2)

Listing 7.2 DFT function.

```
import pylab as pl
```

---

[3] As the DFT is sampled in frequency, each point k refers to a sample, or the complex spectral coefficient at that point.

```
def dft(s):
    N = len(s)
    out = pl.zeros(N)+0j;
    t = pl.arange(0,N)
    for k in range(0,N):
        out[k] = pl.sum(s*(pl.cos(2*pl.pi*k*t/N)
                        - 1j*pl.sin(2*pl.pi*k*t/N)))
    return out/N
```

The DFT is reasonably compact and with Python NumPy arrays, it is fairly simple to define it as a straight translation of Eq. 7.12. The j tag indicates the imaginary part of a complex number. Effectively, there are two loops: one (implicit) performed by sum() over the time t, where we multiply each sample of the input by the complex sinusoid and sum the results; and the other (explicit) over k to produce the N complex coefficients. We can apply this to the sawtooth wave that we programmed in Listing 7.1, obtain the spectrum and plot its magnitudes:

Listing 7.3 DFT analysis of sawtooth wave.

```
import pylab as pl

def dft(s):
    N = len(s)
    out = pl.zeros(N)+0j;
    t = pl.arange(0,N)
    for k in range(0,N):
        out[k] = pl.sum(s*(pl.cos(2*pl.pi*k*t/N)
                        - 1j*pl.sin(2*pl.pi*k*t/N)))
    return out/N

T = 100
saw = pl.zeros(T)
t   = pl.arange(0,T)
N = T//2
for k in range(1,N):
    saw += (1/k)*pl.sin(2*pl.pi*k*t/T)
saw *= 2/(pl.pi)

spec = dft(saw)
mags = abs(spec)
scal = max(mags)

pl.figure(figsize=(8,3))
pl.stem(t,mags/scal, 'k-')
pl.ylim(0,1.1)
pl.tight_layout()
```

```
pl.show()
```

In Fig. 7.2, we see the corresponding magnitude spectrum, with the full positive and negative spectrum. As discussed in Sect. 7.1.1, the negative side can be inferred from the positive frequencies. The magnitudes are mirrored at 0 Hz, as shown in the plots, and there are $\frac{N}{2}+1$ pairs of spectral coefficients. They correspond to equally spaced frequencies between 0 and $\frac{f_s}{2}$ (inclusive).

Each one of these frequency points can be considered as a band, channel, or bin. We are effectively analysing one period of a waveform whose harmonics are multiples of $\frac{f_s}{N}$ Hz, the fundamental analysis frequency. Thus, the output of the DFT is composed of bins centred at $f_c(k) = kf_s/N$. The first N/2 of these are on the positive side of the spectrum; $f_c(N/2) = f_s/2$, is the Nyquist frequency; and the last N/2 points are in the negative spectrum because according to the sampling theorem, $f_c = f_c - f_s$ for $f_c \geq f_s$ (and so the Nyquist point is in both the positive and the negative sides of the spectrum). So we can define the centre frequency points as

$$f_c(k) = \begin{cases} k\frac{f_s}{N}, & 0 \leq k \leq N/2 \\ k\frac{f_s}{N} - f_s, & N/2 \leq k < N/2 \end{cases} \tag{7.13}$$

When input signal is perfectly periodic over $\frac{N}{f_s}$ secs, the analysis will be show well-defined partials lying at the bin centre frequencies. This is the case for the sawtooth with 100 harmonics in Listing 7.3. Note in the plot from Fig. 7.2 that the non-negative spectrum occupies the left side of the DFT output frame, and the negative side follows it (from the lowest frequency upwards), as explained previously for the case of purely real signals.

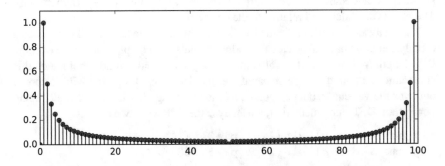

Fig. 7.2 The magnitude spectrum of a band-limited sawtooth wave. The wave period is an integer multiple of the DFT size ($N = 100$), giving a perfect analysis. The full spectrum shows that components appear on the positive and negative sides of the spectrum. The first half of the plot refers to positive frequencies (0 to $\frac{f_s}{2}$), and the second, to negative ones (from $-\frac{f_s}{2}$ to 0 Hz).

As we can observe from the plot in Fig. 7.2, that the positive magnitude spectrum is a mirror image of the negative side. The actual spectral symmetry is a little more subtle. To explore it, let's use the following signal with three components,

$$x(t) = \cos(2\pi 5t/N) + \cos(2\pi 10t/N - \pi/4) + \sin(2\pi 15t/N) \qquad (7.14)$$

namely three harmonics at 5, 10 and $15 f_0$. Note that each harmonic has a different phase: cosine, sine, and the middle one is in between a cosine and a sine (with a phase of $-\pi/4$). The DFT will detect this as follows:

1. Cosine partials show up in the real part of the number.
2. Sine partials appear in the imaginary part.
3. Any component not exactly in one or the other phase will appear as a mixture of the two.

This signal is plotted in Fig. 7.3, where we have four individual graphs. The top two show the real and imaginary spectra. You can see that the first half of the spectrum is symmetric with respect to the second half in the real (cosines) plot, but it is anti-symmetric in the imaginary (sines) plot. This is because cosines are even functions, $\cos(x) = \cos(-x)$, and sines are odd functions, $\sin(x) = -\sin(-x)$ (and the signal we are analysing is purely real). We can also see that the cosine harmonics only show up in the real spectrum, and the sine ones only in the imaginary. The in-between partial is divided between the two. Finally, notice that another feature of the analysis of real input signals is that one partial will always appear on both sides of the spectrum with half the original amplitude. The magnitude plot shows this more clearly. In this form, we are only showing the absolute values, which are a combination of sines and cosines (as per Eq. 7.4). The phase spectrum, from 7.5, shows the exact phases for each partial: 0 (cosine), $\frac{-\pi}{2}$ (sine) and $\frac{-\pi}{4}$ (in between). The phases are calculated relative to the cosine phase.

The analyses we have performed so far have been able to capture all components cleanly, since we have taken care to produce signals that complete one cycle over the DFT length N. However, If we attempt to analyse a signal that does not match this constraint, the output will be smeared, i.e. the detection of a partial will be spread out over the various frequency points, as shown in Fig. 7.4. In this example, using our Python DFT implementation, we analyse the following two signals:

```
sig_perfect = pl.sin(2*pl.pi*3*t/T)
sig_smeared = pl.sin(2*pl.pi*3.333*t/T)
```

The smearing happens because the DFT always represents the segment being analysed as if it were periodic, with period $\frac{N}{f_s}$. The input is always modelled as harmonics of the fundamental frequency of analysis.

The inverse DFT (IDFT) allows us to recover the original signal (whether or not the analysis was perfect). It can be defined as

$$x(t) = \sum_{t=0}^{N-1} X(k) \left[\cos(2\pi kt/N) + j\sin(2\pi kt/N) \right] \qquad k = 0, 1, ..., N-1 \qquad (7.15)$$

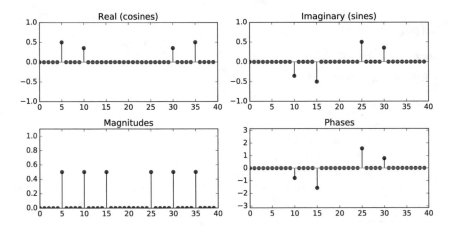

Fig. 7.3 The DFT plots of a signal with three components $x(t) = \cos(2\pi 5t/N) + \cos(2\pi 10t/N - \pi/4) + \sin(2\pi 15t/N)$, showing the separate real and imaginary parts (top), and the magnitudes and phases (bottom), with $N = 40$.

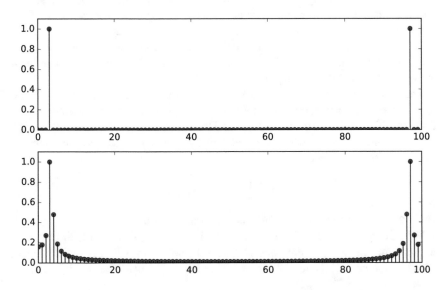

Fig. 7.4 The DFT plots of a sine wave $\sin(2\pi 3t/N)$ (top) and $\sin(2\pi 3.333t/N)$, showing a perfectly-captured partial and a smeared analysis, respectively ($N = 100$).

We can see that this is, in essence an additive synthesis process. In Python, we can program it as:

Listing 7.4 Inverse DFT function.

```python
import pylab as pl

def idft(s):
    N = len(s)
    out = pl.zeros(N);
    k = pl.arange(0,N)
    for t in range(0,N):
        out[t] = pl.sum(s*(pl.cos(2*pl.pi*k*t/N)
                        + 1j*pl.sin(2*pl.pi*k*t/N)))
    return out.real
```

Note that because the original input to the DFT was real, we can assume that the IDFT output will also be purely real (the imaginary part is zero). With this we can verify that $\sin(2\pi 3.333t/N)$ can be analysed and resynthesised correctly.

As we have, smearing affects the clarity of the analysis data. This is related to discontinuities that occur at the ends of the analysis segment, also called the DFT *frame* or *window*. The use of the DFT implies that we have cut one cycle out of an underlying periodic waveform, which corresponds to applying a rectangular envelope to this assumed signal [30]. This is called a *rectangular* window. We can see it applied to the $\sin(2\pi 3.333t/N)$ case in Fig. 7.5.

The *windowing* operation [43] is this application of an envelope to the signal. To mitigate the smearing problem, we can choose a window shape that smooths out discontinuities at the start and end of the DFT frame. An example of such shape is the von Hann window, which we have already encountered in previous chapters. This is a raised inverted cosine, defined as

$$w(t) = \frac{1}{2}[1 - \cos(2\pi t/N)] \tag{7.16}$$

whose application to a sinusoid input is shown in Fig. 7.6.

We can compare the rectangular and von Hann windows to see their effect. In the first case, smearing is present throughout the spectrum, whereas in the second, there is some lateral spread, but otherwise components outside the central band are suppressed (Fig. 7.7).

There are a number of other windows, with different shapes and spectral characteristics, which can be employed for smoothing an input. However, in this chapter, we will use the von Hann window in most applications.

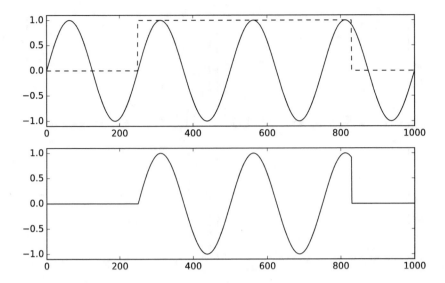

Fig. 7.5 Application of a rectangular window (size N) to a sinusoid (top), selecting an input corresponding to $\sin(2\pi 3.333t/N)$.

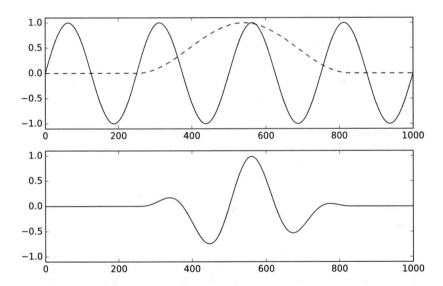

Fig. 7.6 Application of a von Hann window (size N) to a sinusoid (top), selecting an input corresponding to $\sin(2\pi 3.333t/N)$.

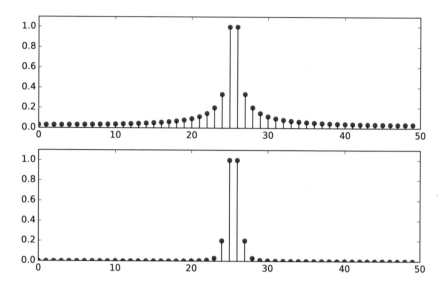

Fig. 7.7 DFT plots (non-negative spectrum only, $N = 100$) of the same signal, where a single partial falls exactly midway between bins 25 and 26, with rectangular (top) and von Hann (bottom) windows.

7.1.3 The fast Fourier transform

The DFT equation (7.12) is not normally implemented in the direct form of Listing 7.2. This is because there are fast algorithms that are much more efficient for computing the corresponding result. These are collectively called the *fast Fourier transform* (FFT), but are effectively the DFT (and not the FT as implied by the name). These algorithms explore symmetries that are inherent in the DFT in order to save computation.

There is a lot of redundancy in the direct implementation of the DFT formula. This can be demonstrated if we state it as a matrix operation [98]. We can do this by setting up an $N \times N$ matrix S containing, in each row, the sinusoidal multipliers, defined by

$$S_{k,t} = \cos(2\pi kt/N) - j\sin(2\pi kt/N) \tag{7.17}$$

Then, with the input signal $x(t)$ in a row vector x, whose components are $x_t = x(t)$, we have

$$X = Sx^T \tag{7.18}$$

with the column vector X containing the spectrum $X(k)$ $(= X_k)$.

Now, as an example, using Euler's formula $e^{j\omega} = \cos(\omega) + j\sin(\omega)$ and $\omega_{k,t} = -2\pi kt/N$, the 4-point DFT can be expressed as

$$
\begin{bmatrix}
e^{j\omega_{0,0}} & e^{j\omega_{0,1}} & e^{j\omega_{0,2}} & e^{j\omega_{0,3}} \\
e^{j\omega_{1,0}} & e^{j\omega_{1,1}} & e^{j\omega_{1,2}} & e^{j\omega_{1,3}} \\
e^{j\omega_{2,0}} & e^{j\omega_{2,1}} & e^{j\omega_{2,2}} & e^{j\omega_{2,3}} \\
e^{j\omega_{3,0}} & e^{j\omega_{3,1}} & e^{j\omega_{3,2}} & e^{j\omega_{3,3}}
\end{bmatrix}
\times
\begin{bmatrix}
x(0) \\ x(1) \\ x(2) \\ x(3)
\end{bmatrix}
$$
$$
=
\begin{bmatrix}
x(0) + x(1) + x(2) + x(3) \\
x(0) + jx(1) - x(2) - jx(3) \\
x(0) - x(1) + x(2) - x(3) \\
x(0) - jx(1) - x(2) + jx(3)
\end{bmatrix}
\tag{7.19}
$$

If we rearrange the sums in the output vector, we have

$$
X =
\begin{bmatrix}
x(0) + x(2) + x(1) + x(3) \\
x(0) - x(2) + jx(1) - jx(3) \\
x(0) + x(2) - x(1) - x(3) \\
x(0) - x(2) - jx(1) + jx(3)
\end{bmatrix}
=
\begin{bmatrix}
a + c \\
b + jd \\
a - c \\
b - jd
\end{bmatrix}
\tag{7.20}
$$

with $a = x(0) + x(2)$, $b = x(0) - x(2)$, $c = x(1) + x(3)$, and $d = x(1) - x(3)$.

This demonstrates how we might be able to factorise the computation to avoid duplicate operations. In this particular case, 16 operations can be reduced to 8, and we have the basic motivation for the various FFT algorithms.

The most common of the FFT algorithms is designed to work with DFT sizes such as the one in Eq. 7.19, which are an exact power-of-two number; this algorithm is known as the *radix-2* FFT. The principle behind it is to split the computation into two half-size transforms, and then repeat this until we have a size-2 DFT [18, 72, 98]. Let's develop the idea so that we can provide an implementation.

Starting with the DFT of Eq. 7.12, but excluding the scaling term,

$$
X(k) = \sum_{t=0}^{N-1} x(t) \left[\cos(2\pi kt/N) - j\sin(2\pi kt/N) \right] = \sum_{t=0}^{N-1} x(t) e^{-2\pi jkt/N}
\tag{7.21}
$$

We proceed to divide this into two parts: one for the even input samples and the other for the odd ones. Each transform is now half the original size:

$$
X_N(k) = \sum_{m=0}^{N/2-1} x(2m) e^{-2\pi jk2m/N} + \sum_{m=0}^{N/2-1} x(2m+1) e^{-2\pi jk(2m+1)/N}
$$
$$
= \sum_{m=0}^{N/2-1} x(2m) e^{-2\pi jkm/(N/2)} + e^{-2\pi jk/N} \sum_{m=0}^{N/2-1} x(2m+1) e^{-2\pi jkm/(N/2)}
$$
$$
= E(k) + e^{-2\pi jk/N} O(k)
$$

$$
\tag{7.22}
$$

with $t = 2m$. The even samples are indexed by $2m$ and the odd ones by $2m + 1$. Thus we have two half-size transforms, $E(k)$ and $O(k)$, acting on the even and odd samples, respectively. The only difference is the $e^{-2\pi jk/N}$ factor, also known as the *twiddle* factor, which multiplies the odd part.

To loop over all frequencies from 0 to k, we use the fact that the DFT is periodic over its length N [4],

$$X(k+N) = X(k) \tag{7.23}$$

which translates to $E(k+N/2) = E(k)$ and $O(k+N/2) = O(k)$ [5]. This leaves us with the following expression for the DFT (for $k = 0,...,N-1$):

$$X_N(k) = \begin{cases} E(k) + e^{-2\pi jk/N}O(k), & 0 \le k < N/2 \\ E(k-N/2) + e^{-2\pi jk/N}O(k-N/2), & N/2 \le k < N \end{cases} \tag{7.24}$$

Using the identity $e^{-2\pi j(k+N/2)/N} = -e^{-2\pi jk/N}$, we can calculate the DFT as a pair of parallel equations, for $k = 0,...,N/2$:

$$\begin{aligned} X_N(k) &= E(k) + \omega^{-k}O(k) \\ X_N(k+N/2) &= E(k) - \omega^{-k}O(k) \end{aligned} \tag{7.25}$$

with $\omega^{-k} = e^{-2\pi jk/N}$.

We can use this principle to split the DFT recursively down to $N = 2$, where Eq. 7.25 can be used with single points, as in

$$\begin{aligned} X_2(0) &= x(0) + x(1) \\ X_2(1) &= x(0) - x(1) \end{aligned} \tag{7.26}$$

To implement the DFT, however, we use loops instead of an explicitly recursive approach. In this case, we nest three loops to achieve an equivalent form to compute the FFT algorithm:

1. The outer loop controls the successive DFT sizes, starting from single points ($N = 2$) and doubling the length at each iteration.
2. The inner loop provides the iterations necessary to process the whole DFT frame, starting with $N_{DFT}/2$ for $N = 2$ and then halving as the individual DFTs double in size (where N_{DFT} is the original DFT length).
3. The middle loop has two functions: to provide the start indices for the inner arithmetic operations and to update the twiddle factor.

---

[4] The underlying assumption is that the DFT is analysing one cycle of a periodic waveform consisting of N samples, and thus the spectrum is also periodic over N.

[5] $\frac{N}{2}$ is the size of each half transform.

We place Eq. 7.25 in the inner loop. This is the business end of the algorithm, and processes pairs of numbers as it progresses through the DFT frame. In the first iteration, it transforms samples that are side-by-side in the frame. In the second, pairs are made of every other sample, then every four samples, and so on, until it processes pairs separated by $\frac{N_{DFT}}{2}$. This iterative process for N_{DFT} is shown in Fig. 7.8, where we can see that the computational complexity involves N operations (columns) performed $\log_2 N$ times (as opposed to N^2 operations for the raw DFT). To improve efficiency, the twiddle factor is calculated as a complex number that is updated by an increment dependent on k and N. The Python implementation is shown in Listing 7.5

Listing 7.5 Radix-2 FFT function.

```python
import pylab as pl

def fft(s):
    N = len(s)
    sig = reorder(s) + 0j
    l2N = int(pl.log2(N))
    for n in [2**x for x in range(0,l2N)]:
        o = -pl.pi/n
        wp = pl.cos(o) + pl.sin(o)*1j
        w = 1+0j
        for m in range(0, n):
            for k in range(m,N,n*2):
                i = k + n
                even, odd = sig[k], w*sig[i]
                sig[k] = even + odd
                sig[i] = even - odd
            w *= wp
    return sig/N
```

In this example, N is N_{DFT}, n takes the place of N, even and odd are $E(k)$ and $O(k)$, respectively, and w is ω. This form of the radix-2 FFT requires the input data points to be reordered in an even-odd arrangement all the way down to each pair of numbers. This allows the internal loop to iterate over the frame and apply Eq. 7.25 to the data. Thankfully, this rearrangement can be understood with minimal difficulty as it is based on a bit-reversal of the frame indices. For example, with $N_{DFT} = 8$, the rearrangement is as follows:

- 000, no change (0)
- 001 → 100, (1 → 4)
- 010, no change (2)
- 011 → 110, (3 → 6)
- 100 → 001, (4 → 1)
- 101, no change (5)
- 110 → 011, (6 → 3)

- 111, no change (7)

The indices (0, 1, 2, 3, 4 , 5, 6, 7) become (0, 4, 2, 6, 1, 5, 3, 7), and we see that all the odd indices are placed in the second half of the frame. This is done similarly for other DFT sizes. The function `reorder()` used in the FFT code rearranges the data in such an order. Since the operation employs complex arithmetic, we make it explicit that the input signal employs a complex type by adding `0j` to the re-ordered output. This allows both real and complex inputs to be used with this function.

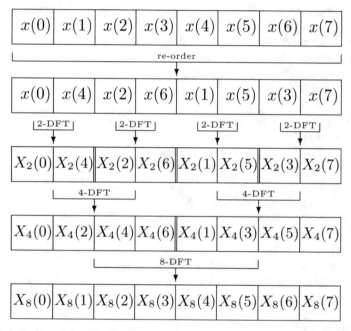

Fig. 7.8 A diagram of the radix-2 FFT non-recursive algorithm, with $N_{DFT} = 8$.

A full example using the additive-synthesis sawtooth wave as input is shown in Listing 7.6. This demonstrates that this FFT algorithm is equivalent to the plain DFT, although it is limited to power-of-two sizes. The spectrum plot produced by this program is shown in Fig. 7.9.

Listing 7.6 FFT analysis of a sawtooth wave.

```
import pylab as pl


def reorder(s):
    N = len(s)
    o = s.copy()
    j = 0
    for i in range(0,N):
```

```
        if j > i:
          o[i] = s[j]
          o[j] = s[i]
        m = N//2
        while m >= 2 and j >= m:
          j -= m
          m = m//2
        j += m
    return o


def fft(s):
    N = len(s)
    sig = reorder(s) + 0j
    l2N = int(pl.log2(N))
    for n in [2**x for x in range(0,l2N)]:
      o = -pl.pi/n
      wp = pl.cos(o) + pl.sin(o)*1j
      w = 1+0j
      for m in range(0, n):
        for k in range(m,N,n*2):
          i = k + n
          even, odd = sig[k], w*sig[i]
          sig[k] = even + odd
          sig[i] = even - odd
        w *= wp
    return sig/N


T = 128
saw = pl.zeros(T)
t   = pl.arange(0,T)
N = T//2
for k in range(1,N):
    saw += (1/k)*pl.sin(2*pl.pi*k*t/T)
saw *= 2/(pl.pi)

spec = fft(saw)
mags = abs(spec)
scal = max(mags)

pl.figure(figsize=(8,3))
pl.stem(t,(mags/scal), 'k-')
pl.ylim(0,1.1)
pl.xlim(0,T)

pl.tight_layout()
```

```
pl.show()
```

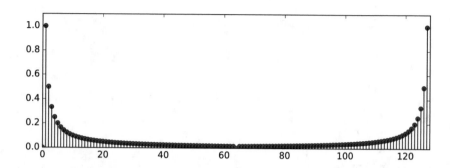

Fig. 7.9 The magnitude spectrum of a band-limited sawtooth wave as computed by an FFT algorithm, with $N = 128$.

Finally, to implement the inverse FFT, we can use the exact same algorithm, since the process is identical. To make sure it matches Eq. 7.15, though, we need to adjust the twiddle factor update by changing its sign and removing the output scaling factor. Also, because the input is complex, we do not need to convert it explicitly as we did in the case of the forward transform (Listing 7.7).

Listing 7.7 Radix-2 IFFT function.

```python
import pylab as pl

def ifft(s):
    N = len(s)
    sig = reorder(s)
    l2N = int(pl.log2(N))
    for n in [2**x for x in range(0,l2N)]:
        o = pl.pi/n
        wp = pl.cos(o) + pl.sin(o)*1j
        w = 1+0j
        for m in range(0, n):
            for k in range(m,N,n*2):
                i = k + n
                even, odd = sig[k], w*sig[i]
                sig[k] = even + odd
                sig[i] = even - odd
            w *= wp
    return sig
```

It is possible to verify that the analysis process is fully transparent by writing out = ifft(fft(sig)) and comparing the input and output.

In audio applications, we will most likely be using the DFT to process purely real data. Because of this, we can ignore the negative half of the spectrum, as we have already seen. However, it is also possible to exploit this redundancy to improve the computation of the FFT one step further [68]. More formally, the DFT symmetry means that, for real signals, the negative spectrum is the complex conjugate of the positive side (and vice versa):

$$X(k) = \overline{X(N-k)} \tag{7.27}$$

where $\overline{a+jb} = a - jb$.

By taking advantage of this, we can pack twice the data into a single transform [16, 72]. For N real-valued inputs, we need only an $\frac{N}{2}$ complex DFT. In order to do this, we need to work out how to get the correct results from a half-sized DFT. We begin by recalling Eq. 7.22, since the process has similarities to the principles of the radix-2 FFT.

$$
\begin{aligned}
X(k) &= \sum_{t=0}^{N-1} x(t) e^{-2\pi jkt/N} \\
&= \sum_{m=0}^{N/2-1} x(2m) e^{-2\pi jkm/(N/2)} + e^{-2\pi jk/N} \sum_{m=0}^{N/2-1} x(2m+1) e^{-2\pi jkm/(N/2)}
\end{aligned}
\tag{7.28}
$$

Now we take $x(2m) = a(m)$ and $x(2m+1) = b(m)$ to be the real and imaginary parts of $y(m) = a(m) + jb(m)$. The DFT of $y(t)$ is

$$
\begin{aligned}
Y(k) &= \sum_{m=0}^{N/2-1} y(m) e^{-2\pi jkm/(N/2)} = \sum_{m=0}^{N/2-1} [\Re\{y(m)\} + j\Im\{y(m)\}] e^{-2\pi jkm/(N/2)} \\
&= \sum_{m=0}^{N/2-1} \Re\{y(m)\} e^{-2\pi jkm/(N/2)} + j \sum_{m=0}^{N/2-1} \Im\{y(m)\} e^{-2\pi jkm/(N/2)} \\
&= R(k) + jI(k)
\end{aligned}
\tag{7.29}
$$

We have managed to re-interpret the real input as a complex signal, and then split it into two DFTs, $R(k)$ and $I(k)$, of the separate real and imaginary parts, respectively. These are two transforms of real signals, so we have the same symmetry relations between the two halves of their spectra, as per Eq. 7.27:

$$R(k) = \overline{R(N-k)} \tag{7.30}$$

$$I(k) = \overline{I(N-k)} \tag{7.31}$$

Furthermore, we can write the complex conjugate of $Y(k)$ in terms of $R(k)$ and $I(k)$:

$$\overline{Y(N-k)} = \overline{R(N-k)} + \overline{jI(N-k)} = R(k) - jI(k) \tag{7.32}$$

from eqs. 7.29 and 7.32, we have

$$R(k) = \frac{1}{2}\left(Y(k) + \overline{Y(N/2-k)}\right) \tag{7.33}$$

$$I(k) = \frac{j}{2}\left(\overline{Y(N/2-k)} - Y(k)\right) \tag{7.34}$$

Substituting these results back into Eq. 7.28 we have:

$$X(k) = \sum_{m=0}^{N/2-1} a(m)e^{-2\pi jkm/(N/2)} + e^{-2\pi jk/N}\sum_{m=0}^{N/2-1} b(m)e^{-2\pi jkm/(N/2)} \tag{7.35}$$
$$= R(k) + \omega^{-k}I(k)$$

for $k = 0,1,2,...,N/2-1$, with $\omega^{-k} = e^{-2\pi jk/N}$. To get the other side of the spectrum, $k = N/2+1,...,N-1$, we can use eqs. 7.30 and 7.31 over the same range $0 \le k < N/2$. Taking $e^{-2\pi j(k+N/2)/N} = -e^{-2\pi jk/N}$, as we did in Eq. 7.25, we have $\omega^{-(N-k)} = -\omega^{-k}$, and so the expression for points $N/2+1$ to $N-1$ becomes

$$X(N-k) = \overline{R(k) - \omega^{-k}I(k)} \tag{7.36}$$

The final point $X(N/2)$ is just $R(0) - I(0)$, which together with eqs. 7.35 and 7.36 gives us all we need to write a conversion function exploiting this principle. The code will follow these equations quite closely; here's the outline:

1. Reinterpret the length-N real input $x(t)$ as a length-$N/2$ complex signal $y(m) = x(2m) + jx(2m+1)$ by placing every other number into a real or an imaginary part.
2. Apply the ordinary complex DFT (using the FFT algorithm), producing $N/2$ complex numbers.
3. Convert the 0 Hz and Nyquist points and store them in the real and imaginary parts of the first point in the output frame. This can be done because these two points are purely real (given that $\sin(0)$ and $\sin(\pi t)$ in the DFT equation are both zero and the input signal itself is real-valued).
4. Loop over the range 1 to $N/2$ and apply eqs. 7.33, 7.34, 7.35 and 7.36.

The function `rfft()` is shown below as part of a complete program that analyses and plots a band-limited sawtooth wave (Listing 7.8).

Listing 7.8 FFT analysis of a sawtooth wave, optimised for real inputs.

```
import pylab as pl

def reorder(s):
    N = len(s)
    o = s.copy()
```

```
      j = 0
      for i in range(0,N):
        if j > i:
          o[i] = s[j]
          o[j] = s[i]
        m = N//2
        while m >= 2 and j >= m:
          j -= m
          m = m//2
        j += m
      return o


def fft(s):
    N = len(s)
    sig = reorder(s) + 0j
    l2N = int(pl.log2(N))
    for n in [2**x for x in range(0,l2N)]:
      o = pl.pi/n
      wp = -pl.cos(o) + pl.sin(o)*1j
      w = 1+0j
      for m in range(0, n):
        for k in range(m,N,n*2):
          i = k + n
          even, odd = sig[k], w*sig[i]
          sig[k] = even + odd
          sig[i] = even - odd
        w *= wp
    return sig/N


def rfft(s):
    N = len(s)
    sig = s[:N:2] + s[1:N:2]*1j
    s = fft(sig)
    N = N//2
    o = -pl.pi/N
    w = complex(1,0)
    R = s[0].real*0.5
    I = s[0].imag*0.5
    zero =  R + I
    nyqs =  R - I
    s[0] = zero.real + 1j*nyqs.real
    wp = pl.cos(o) + pl.sin(o)*1j
    w *= wp
    for i in range(1,N//2+1):
        j = N - i
```

```
        R = 0.5*(s[i] + s[j].conjugate())
        I = 0.5j*(s[j].conjugate() - s[i])
        s[i] = R + w*I
        s[j] = (R - w*I).conjugate()
        w *= wp
    return s


T = 128
saw = pl.zeros(T)
t   = pl.arange(0,T)
N = T//2
for k in range(1,N):
    saw += (1/k)*pl.sin(2*pl.pi*k*t/T)
saw *= 2/(pl.pi)

spec = rfft(saw)
mags = abs(spec)
scal = max(mags)

pl.figure(figsize=(8,3))
pl.stem(t[0:N],mags/scal, 'k-')
pl.ylim(0,1.1)
pl.xlim(0,N)

pl.tight_layout()
pl.show()
```

A plot generated by the optimised FFT analysis of a real input is shown in Fig. 7.10. Note that we now only have the non-negative spectrum to plot, $N/2$ points. A small detail is that if we want to plot the magnitude of the Nyquist frequency, we will need to retrieve it from the imaginary part of the first point (in the present case both 0 Hz and the Nyquist frequency magnitudes are 0).

To complement this discussion, we have the inverse operation, which recovers the spectrum of an $\frac{M}{2}$ complex DFT with just some small modifications to the equations. It then applies `ifft()` to it and interleaves the real and imaginary parts of the DFT output to obtain a length-N real waveform, $x(t) = \Re\{y(\lfloor t/2 \rfloor)\}(1 - t \bmod 2) + \Im\{y(\lfloor t/2 \rfloor)\}(t \bmod 2)$:

Listing 7.9 IFFT function, optimised for spectra of real signals.

```
import pylab as pl


def irfft(s):
    N = len(s)
    w = complex(1,0)
    nyqs = s[0].imag*2.
    zero = s[0].real*2.
```

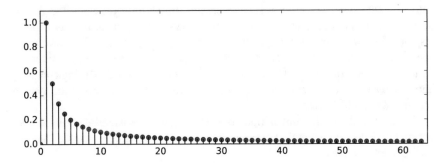

Fig. 7.10 The magnitude spectrum of a band-limited sawtooth wave computed with the optimised *real* FFT algorithm, with $N = 128$, yielding 64 frequency points.

```
R = (zero + nyqs)
I = 1j*(zero - nyqs)
s[0] =   R + w*I
o = pl.pi/N
wp = pl.cos(o) + pl.sin(o)*1j
w *= wp
for i in range(1,N//2+1):
    j = N - i
    R = 0.5*(s[i] + s[j].conjugate())
    I = 0.5j*(s[i] - s[j].conjugate())
    s[i] = R + w*I
    s[j] = (R - w*I).conjugate()
    w *= wp
s = ifft(s)
sig = pl.zeros(2*N)
sig[0::2] = s.real
sig[1::2] = s.imag
return sig
```

This pair of functions is often known as the *real* FFT and *real* IFFT. It is important to point out, however, that both work with complex numbers (the FFT produces and the IFFT consumes them) as usual, and so we must not be confused by the terminology.

7.1.4 Convolution reverb

One of the typical applications of the DFT (FFT) is in the implementation of *convolution reverb*. This is used to impart the characteristics of a space, such as room, onto an input sound. The process is conceptually straightforward: we collect the reflective qualities of the system in question (e.g. the space) by measuring its impulse response (IR), and then we combine this with the input signal using the convolution operation. We have already seen the concept of an IR when we looked at filters of finite and infinite IR types, where we defined it as their response to a single-sample impulse. The concept is similar here, and the measured IR will make the system equivalent to an FIR filter.

We can now define the convolution operation and see how it can be implemented. For this, let's look at a simple averaging filter, which calculates a weighted sum of two adjacent samples:

$$y(t) = \frac{1}{2}x(t) + \frac{1}{2}x(t-1) \tag{7.37}$$

This is equivalent to applying an IR consisting of two samples, $h(t) = [0.5, 0.5]$, to an input signal $x(t)$ via a convolution operation. Examining the process more closely we see that $x(t)$ is delayed by 0 and 1 samples, producing two streams. Each one of these is scaled by the corresponding IR sample $(h(0), h(1))$, and these are subsequently mixed together. These are the steps in convolution: delay, scale and mix. More formally, we have, for two finite-length signals,

$$y(t) = x(t) * h(t) = \sum_{n=0}^{N-1} x(t-n)h(n) \tag{7.38}$$

where N is the length of the IR ($N = 2$) in the example above. The output signal will have a length $N + L - 1$, where L is the length of the input.

To apply this concept to reverberation, we should think of the measured IR as a collection of reflection amplitudes (weights) at different delay times (to a resolution of $\frac{1}{f_s}$). The convolution is a mix of all of these scaling values applied to delayed copies of the input signal. Thus, to implement this, a long delay line of size N tapped at each sample, with an associated gain given by the IR, should be sufficient. This is called *direct*, or delay-line, convolution.

This method would indeed be all we needed, if it were not for the heavy computational load it imposes. When we attempt to implement this, we realise that, for each sample of the output, we need N delay line lookups, N multiplications and a sum of N samples. With moderate IR lengths, this will perform poorly, and with large ones it will not be practical. For small IRs, however, we can use this method.

The solution is to use the following relation:

$$x(t) * h(t) = X(k) \times H(k) \tag{7.39}$$

where $X(k)$ and $H(k)$ are the spectra of $x(t)$ and $h(t)$. This allows us to use the FFT to implement the convolution of two signals.

There a few practical issues that we will have to deal with to make this ready for general use. First of all, the length L of the input is likely to be much larger than that of the IR, N, and in some cases (e.g. real-time audio), it will not be known. The expression in Eq. 7.39 requires that we know what L is. If we do, then we can apply the expression directly:

$$y(t) = IDFT_M(DFT_M(x(t)) \times DFT_M(h(t))) \qquad (7.40)$$

where M needs to be at least $N + L - 1$. To use the FFT, we need to set M to a power of two that is equal to or greater than this. In this case, we have what is known as *fast* convolution. The inputs to the FFT will be smaller than this, but we can insert zeros to complete each signal (*padding* by zeros).

This will work if we do not care about waiting until all the samples have been collected for the FFT. Clearly this will not be appropriate for real-time applications, as we need to know the input length beforehand, and in any case we cannot wait for the input to be filled up. It might also be problematic with very large input lengths,

A more practical solution is to use the length N of the IR as the basis for the convolution [42, 68], producing a series of partial convolution outputs $Y_n(k)$:

$$Y_n(k) = X_n(k)H(k) \qquad (7.41)$$

where

$$X_n(k) = DFT_M(w(t)x(t+nN)) \qquad 0 < n < \lceil L/N \rceil \qquad (7.42)$$

with $w(t)$ being a rectangular window of size N and M being the DFT length.

In this case, we set M to the next power of two such that $M > 2N - 1$, then slice the input into $\lceil L/N \rceil$ segments[6], and apply Eq. 7.40 to zero-padded data. The convolution output is then defined by

$$y(t) = \sum_{n=0}^{\lceil L/N \rceil - 1} y_n(t) \qquad (7.43)$$

where $y_n(t) = w(t - nN)IDFT(Y_n(k))$, with $w(t)$ a rectangular window of size $2N - 1$. The partial convolutions are added together to reconstruct the signal, each starting at the nN boundary from where it was extracted originally. Since these slices are each $2N - 1$ in length, they will overlap by $N - 1$ samples, making Eq. 7.43 define a process called *overlap-add* (Fig. 7.11).

This is implemented as a Python function in Listing 7.10. A full program using this function is shown in Appendix B (sect. B.6). From now on, we will use the natively-implemented FFT functions from the pylab package, as they are more efficient than the pure-Python versions studied in Sect. 7.1.3. For all practical purposes, they are equivalent, except for the fact that while we were packing the Nyquist frequency in point 0 in our version, the pylab functions use an extra point in its output array ($N/2 + 1$ complex numbers). Another small difference is that scaling by $1/N$

---

[6] The expression $y = \lceil x \rceil$ indicates an integer $y \geq x$. The last input segment will be padded with zeros if L/N is not integral.

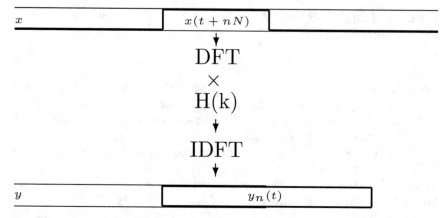

Fig. 7.11 Fast convolution diagram showing windowed input, spectral processing and overlap-add output.

happens in the inverse stage, rather than in the forward transform as was the case with the code we developed earlier on[7].

Listing 7.10 Overlap-add FFT-based convolution.

```
import pylab as pl

def conv(signal,ir):
    N = len(ir)
    L = len(signal)
    M = 2
    while(M <= 2*N-1): M *= 2

    h = pl.zeros(M)
    x = pl.zeros(M)
    y = pl.zeros(L+N-1)

    h[0:N] = ir
    H = pl.rfft(h)
    n = N
    for p in range(0,L,N):
     if p+n > L:
            n = L-p
            x[n:] = pl.zeros(M-n)
     x[0:n] = signal[p:p+n]
     y[p:p+2*n-1] += pl.irfft(H*pl.rfft(x))[0:2*n-1]
```

---

[7] In many cases, it is more convenient to have the scaling in the forward transform, as it gives a cleaner set of magnitudes that are not scaled up by the DFT size. However, many FFT implementations, such as the one in pylab, place it in the inverse transform.

`return` y

This implementation of convolution generally works very well, but it has one issue that prevents it from being useful for real-time applications: if the impulse response is very long, there is a noticeable delay between input and output. This is the same problem as before; we have to wait until we have collected enough samples to do the FFT before we can process and output them[8].

To deal with this, we can extend the principle of segmentation [99], so that now we break down the IR into smaller sections, which allows us to start outputting as soon as the first segment has been processed [103]. This is known as *partitioned* convolution and it is the basic algorithm used in most applications of the technique. In this case, we have the following expression:

$$Y_n(k) = \sum_{m=0}^{\lceil N/S \rceil - 1} X_{n-m}(k)H_m(k) \tag{7.44}$$

where $H_m(k) = DFT(w(t)h(t+mS))$, the partitions of the impulse response are $h(t)$, and S is the partition size. This reduces to Eq. 7.38 for $S = 1$.

The general outline of partitioned convolution is given by the following steps (Fig. 7.12):

1. Partition the IR into $P = \lceil N/S \rceil$ partitions of S (power-of-two) length.
2. Take P 2S-DFT analyses of the IR data and store them, reverse-ordered, in an array.
3. For each S segment of the input data, take a zero-padded 2S-DFT analysis and place it in a circular buffer.
4. P complex multiplications of the last P inputs and the IR segments are summed together to derive the current output frame.
5. A $2S$ segment is synthesised using the inverse DFT.
6. The output is composed of the first S samples of the current frame mixed with the final S samples of the previous. We save the second half of the current DFT output to use next time.

We are overlapping the first half output with the saved samples of the previous analysis, in contrast to the earlier version of overlap-add we saw in the single-partition method. The central element of partitioned convolution is the use of a spectral-domain delay line, into which we place the input DFT frames before multiplying them by the impulse response, as shown in Fig. 7.12. This is similar to direct convolution, except that instead of single samples, we have DFT blocks in the delay line.

A Python function implementing this method is shown in Listing 7.11. It can be used as a drop-in replacement for the one in Listing 7.10 and in the Appendix example (Sect. B.6).

---

[8] Note that in offline-processing cases, such as the examples in Appendix B, this is not an issue as we do not have to produce and consume samples in real time.

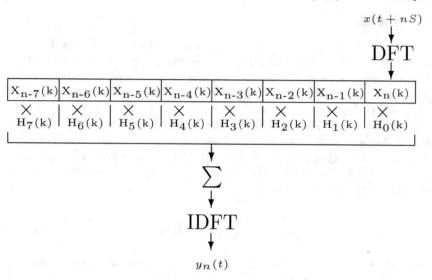

Fig. 7.12 Partitioned convolution diagram (8 partitions).

Listing 7.11 Partitioned convolution.

```
import pylab as pl

def pconv(signal,ir,S=1024):
    N = len(ir)
    L = len(signal)
    M = S*2

    P = int(pl.ceil(N//S))
    H = pl.zeros((P,M//2+1)) + 0j
    X = pl.zeros((P,M//2+1)) + 0j
    x = pl.zeros(M)
    o = pl.zeros(M)
    s = pl.zeros(S)

    for i in range(0,P):
        p = (P-1-i)*S
        ll = len(ir[p:p+S])
        x[:ll] = ir[p:p+S]
        H[i] = pl.rfft(x)

    y = pl.zeros(L+N-1)
    n = S
    for p in range(0,L+N-S,S):
```

```
if p > L:
    x = pl.zeros(M)
else:
  if n+p > L:
      n = L - p
      x[n:] = pl.zeros(M-n)
   x[:n] = signal[p:p+n]
X[i] = pl.rfft(x)
i = (i+1)%P
O = pl.zeros(M//2+1) + 0j
for j in range(0,P):
    k = (j+i)%P
    O += H[j]*X[k]
o = pl.irfft(O)
y[p:p+S] = o[:S] + s
s = o[S:]

return y
```

The partitioned convolution method is a compromise between the demanding direct method and the high-latency single-partition fast convolution. By reducing the partition size, we transfer more computation to the delay-line multiplication loop, reducing the latency, but increasing the processor load. Notice that Eq. 7.44 becomes Eq. 7.39, if $S = N$ and Eq. 7.38 if $S = 1$ (with the identities $X(k) = x(t)$ and $H(k) = h(t)$ since the transform size is 1). In fact, the function in listing 7.11 can also be applied in such cases. An example of the application of this algorithm is shown in Appendix B (Listing B.17)

It is possible to use multiple partition sizes (*non-uniform* partitioned convolution) to achieve a balance between low-delay (or no delay) and good performance [27]. In this case, we can use direct convolution for the first block of N samples of the IR, then partitioned convolution of N-length for the next $M - N$ samples ($M > N$, e.g. $M = 4N$), then a M-length partition after that, and so on. This will avoid any delays and also allow us to use larger FFT sizes for the tail side of the IR[9]. For these real-time applications, Csound has `dconv` and `ftconv` for direct and partitioned convolution, respectively. Both of them work with function tables holding the impulse response and can be combined together to implement zero-latency non-uniform partitioned convolution. An example of a Csound UDO implementing this method is also presented in Listing B.17 in Appendix B.

Convolution reverb can produce the most realistic results if we have at our disposal a good set of IRs for various types of spaces. For a stereophonic effect, these can be measured with a pair of microphones and applied separately to a source to produce two audio channels. Although we have discussed and implemented monophonic examples, they can be extended to multiple-channel applications. Finally,

---

[9] Note also that using multiple partitions might not be as efficient as using a single partition, and so this method is not recommended for non-real-time applications.

convolution is a general-purpose application and we can use it in different ways, by reinterpreting sounds as IRs. We can also use it to implement FIR filtering, in which case we can derive the impulse response from the IDFT of a frequency response of our choice. This can be a very flexible way to design filters, as discussed for Csound applications in [59].

7.2 Time-Varying Spectral Analysis and Synthesis

The previous sections have dealt with spectral analyses that were, at least conceptually, fixed in time. In order to provide a more flexible way to capture and transform dynamic spectra, we need to develop a version of the DFT that can be dependent on time as well as on frequency. This will be done by introducing the concept of the short-time Fourier transform (STFT), which is based on the principles developed for the DFT.

In fact, we will use some of the ideas already introduced in Sect. 7.1.4, when we segmented the input to the DFT by applying a rectangular windon $w(t)$ in Eq. 7.42. This is the basis of the method that we will work out here. Taking an STFT of a segment extracted by $w(t)$ (now a window of any shape), zero everywhere except for $0 \leq t < N$, starting at time n is equivalent to the following expression [43]:

$$X_n(k) = \sum_{t=n}^{n+N/1} w(t-n)x(t)e^{-2\pi jkt/N} \tag{7.45}$$

This is the basic form of the STFT, a transform localised at time n. Notice that the difference is that now we do not assume anything about the duration of $x(t)$, which can extend forever in time. Because of this, the starting point n for the analysis can be anywhere from $-\infty$ to ∞, and so can t^{10}, an therefore we cannot apply the DFT directly to this because its formula requires the summation index to be between 0 and $N-1$. To achieve this, we replace t by $m+n$ ($m = t-n$), leading the index m to start at 0:

$$\begin{aligned} X_n(k) &= \sum_{m=0}^{N-1} w(m)x(m+n)e^{-2\pi jk(m+n)/N} \\ &= e^{-2\pi jkn/N}\sum_{m=0}^{N-1} w(m)x(m+n)e^{-2\pi jkm/N} \end{aligned} \tag{7.46}$$

---

[10] It should very simple to show that although $-\infty < t < \infty$, the summation can be reduced to $n \leq t < n+N$ as in 7.45. We can for instance, split the summation into a sum of three terms, covering $-\infty < n$, $n \leq t < n+N$, and $n+N < \infty$, where the only non-zero part is the middle one. There are other, more complicated ways to get to this result, but this should be sufficient to justify it.

So we have worked out the STFT in terms of a DFT that is scaled by a complex exponential. This exponential is there to account for the fact that there is a shift in time that introduces a phase offset to the DFT. This expression allows us to calculate the STFT at points $n = 0, l, 2l, 3l, ...$, where l is a time increment called the *hop size*. This can generally be much larger than a single sample, but it depends on factors such as the window shape and the DFT size. If $l = N$, which is the case if we have no overlaps in the analysis, the complex multiplying factor becomes 1 and we can forget about it.

For time-varying analysis applications, however, we should ideally use $l < N$. In this case, we need to take account of this factor in order for the equation to be correctly implemented. The simplest thing to do is to rotate the samples circularly in the input array. Owing to the symmetry of the DFT, this multiplication is equivalent to rotating the samples of the input signal $n \bmod N$:

$$e^{2\pi jkn/N}x(m) = x([m+n] \bmod N) = ROT_n^N(x(m)) \tag{7.47}$$

With this relationship, the STFT can be fully defined in terms of the DFT of a windowed input frame starting at time n:

$$X_n(k) = DFT_N(ROT_{-n}^N[x_n(t)]) \tag{7.48}$$

where $x_n(t) = w(t-n)x(t)$ and ROT_n^N as the rotation operation (Eq. 7.47).

Alternatively, we can multiply the DFT output by $e^{-2\pi jkn/N}$, which is equivalent to shifting the phase of each frequency bin k by $-2\pi kn/N$. In that case, we have the STFT simply as

$$X_n(k) = \omega^{-kn}DFT_N(w(t-n)x(t)) \tag{7.49}$$

with $\omega^{kn} = e^{2\pi jkn/N}$. Depending on what we intend to do with the STFT, we can choose to implement one version or the other[11]. In some applications, where this phase shift is not important, we might actually be able to ignore it. One of these situations is when we do not need the exact phase values and so they can contain an offset. The offset is removed when we apply the inverse operation to get the waveform back.

In order to recover the original signal, we need to retrace the steps above by applying an IDFT to the nth analysis frame, rotating it back (if necessary, or we could multiply the spectrum by ω^{-kn} before the IDFT), and then applying the window. Now, this is where we have to be careful. To actually recover the exact waveform block, we would need to divide the IDFT result by $w(t)$. This is possible if the window does not have zeros anywhere, but if we are using the von Hann window, that will not be the case. There are windows that do have this characteristic (like the rectangular window); one of them is the *Hamming*, which, like the von Hann window, is a raised scaled cosine:

---

[11] The multiplication can also be transformed into a phase offset, if that is convenient. Of the three methods, rotation, phase offset, and complex multiplication, the last one is the least efficient. We implement it here only to reinforce the principles developed in the text.

$$w(t) = 0.54 + 0.46\cos(2\pi t/N) \tag{7.50}$$

However, in most practical applications, we will want to use windows that tend to zero at the ends. Also, if we modify the spectral data, recovering it with the exact inverse process might not lead to good results. So, instead, we apply the window with another multiplication. This will have the desired smoothing effect for processes that require it. The complete signal is then recovered using overlap-add in a similar way to Eq. 7.43:

$$x(t) = \sum_{n=0}^{\lceil L/l\rceil-1} x_n(t) \tag{7.51}$$

where $x_n(t) = w(t - nl)IDFT(\omega^{kn}X_n(k))$, l is the hop size and L is the signal total duration. For most practical applications, we should try to use the same analysis and synthesis hop sizes, which will help us avoid further issues in reconstruction.

We now have a complete time-varying analysis and synthesis method, which we can use to design spectral processing instruments. As we have done before, let's implement these ideas in working Python code. Listing 7.12 implements two functions, stft() and istft(), which are used in a test program, analysing a 50 Hz sine wave. The resulting plot confirms that the analysis-synthesis process is transparent (apart from the fade in and out introduced by the windowing process). The hop size is kept at $N/4$, which is the minimum for full reconstruction with the von Hann window (see Sect. 7.2.3).

Listing 7.12 STFT test program.

```python
import pylab as pl

def stft(x,w,n):
    N = len(w)
    X = pl.rfft(w*x)
    k = pl.arange(0,N/2+1)
    return X*pl.exp(-2*pl.pi*1j*k*n/N)

def istft(X,w,n):
    N = len(w)
    k = pl.arange(0,N/2+1)
    xn = pl.irfft(X*pl.exp(2*pl.pi*1j*k*n/N))
    return xn*w

N = 1024
D = 4
H = N//D

t = pl.arange(0,N)
win = 0.5 - 0.5*pl.cos(2*pl.pi*t/N)
```

```
sr = 44100
f = 50
L = sr
t = pl.arange(0,L)
signal = pl.sin(2*pl.pi*f*t/sr)
output = pl.zeros(L)
scal = 1.5*D/4

for n in range(0,L,H):
    if(L-n < N): break
    frame = stft(signal[n:n+N],win,n)
    output[n:n+N] += istft(frame,win,n)

pl.figure(figsize=(8,3))
pl.ylim(-1.1,1.1)
pl.plot(t/sr,output/scal,'k-')
pl.show()
```

Note that we have implemented the STFT and ISTFT equations including the phase shifts, which in this particular case cancel out. This is one example of where we can omit them, although they have been included here for correctness, since we will need this shift when manipulating frequencies later in section 7.2.3.

7.2.1 Processing spectral data

With the STFT in our hands, we can process spectral data to make different types of audio effects. While we could apply the processing directly to the complex (rectangular) real and imaginary coefficients, it is more intuitive to convert these to magnitudes and phases. This allows us to modify the amplitude spectrum separately, which will be our aim in this section. The conversion can be performed by using the Python functions abs() and angle(), for magnitudes and phases, respectively. While the first one is readily available as a built-in method of the complex type, the second comes from pylab.

This splits the spectrum into two separate real arrays. To convert back into a complex real and imaginary frame, we can use this relationship:

$$X(k) = A(k)\cos(\phi(k)) + jA(k)\sin(\phi(k)) \tag{7.52}$$

with $A(k)$ and $\phi(k)$ standing in for the magnitude and phase spectra, respectively. This translates into the one-line Python function

```
def p2r(mags, phs):
    return mags*pl.cos(phs) + 1j*mags*pl.sin(phs)
```

The first process we can apply is to use an amplitude mask and filter out some bins. For instance, we can write a program to remove high frequencies above a certain cut off. To translate a value f_c in Hz into a bin index k, we use the following relation, where N is the DFT size:

$$k = N \frac{f_c}{f_s} \tag{7.53}$$

This follows from the fact that the spectral frame covers the range 0 to $\frac{f_s}{2}$, with $\frac{N}{2}+1$ points. Listing 7.13 presents a code excerpt that demonstrates this. It uses a mask that is 0.5 from 0 to 2000 Hz, and 0 elsewhere, implementing a low-pass effect.

Listing 7.13 Spectral masking.

```
import pylab as pl

fc = 2000
mask = pl.zeros(N//2+1)
mask[:fc*N//sr] = 0.5

for n in range(0,L,H):
    if(L-n < N): break
    frame = stft(signal[n:n+N],win,n)
    mags = abs(frame)
    phs = pl.angle(frame)
    mags *= mask
    frame = p2r(mags,phs)
    output[n:n+N] += istft(frame,win,n)
```

The mask can be varied on a frame-by-frame basis for dynamic spectral modification. We can also use a less crude mask by shaping it with a smooth curve of some description. For instance, a band-pass filter centred at `fc` can be created with the following code:

```
fc = 1000
w = 500
mask = pl.zeros(N//2+1)
d = int(w*N/sr)
b = int(fc*N/sr)
mask[b-d:b+d] = pl.hanning(2*d)
```

where `hanning()` is a von Hann window shape (Fig. 7.13). We can move this mask around dynamically from a centre at 1000 to one at 5000 Hz, making the process a time-varying band-pass filter (Listing 7.14).

Listing 7.14 Time-varying band-pass filter.

```
fc = 1000
fe = 5000
```

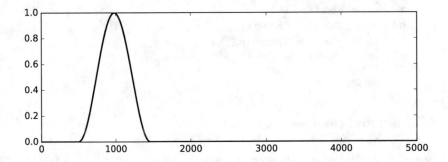

Fig. 7.13 Band-pass spectral mask based on a von Hann window centred at 1000 Hz.

```
incr = (fe - fc)*(H/L)

for n in range(0,L,H):
    if(L-n < N): break
    frame = stft(signal[n:n+N],win,n)
    mags = abs(frame)
    phs = pl.angle(frame)
    mask = pl.zeros(N//2+1)
    d = int((fc/2)*N/sr)
    b = int(fc*N/sr)
    mask[b-d:b+d] = pl.hanning(2*d)
    fc += incr
    mags *= mask
    frame = p2r(mags,phs)
    output[n:n+N] += istft(frame,win,n)
```

A skeleton program for spectral masking is shown in Appendix B (sect. B.7), which can be modified to implement different types of filtering.

We can extend this idea to using the magnitudes of an arbitrary input sound as the mask. In this case we will analyse two inputs, multiply their magnitudes and use the bin phases of one of them. This is a type of cross-synthesis where the common components of the two sounds are emphasised. The bins whose magnitudes are zero or close to zero in one of the sounds are suppressed. The code excerpt in Listing 7.15 demonstrates the process.

Listing 7.15 Cross synthesis example.

```
for n in range(0,L,H):
    if(L-n < N): break
    frame1 = stft(signal1[n:n+N],win,n)
    frame2 = stft(signal2[n:n+N],win,n)
    mags1 = abs(frame1)
```

```
mags2 = abs(frame2)
phs = pl.angle(frame1)
frame = p2r(mags1*mags2,phs)
output[n:n+N] += istft(frame,win,n)
```

7.2.2 Spectral envelope

One of the characteristics of the cross synthesis process outlined above is that it
depends on the existence of common components. In some cases this may be problematic. For instance, say we want to derive a mask that has the overall spectral
shape of a sound, but not the separate partial peaks. For this type of application, we
need to be able to extract the *spectral envelope*, which is a smooth shape linking all
the partial amplitudes (Fig. 7.14).

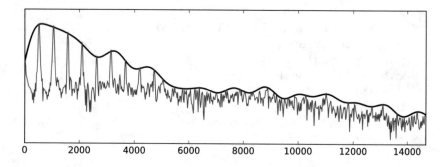

Fig. 7.14 Spectral envelope and the underlying spectrum.

To extract the spectral envelope we need to be able to smooth out the dips in the
magnitude spectrum. For this purpose, we introduce the principle of the cepstrum
transform [10], which is defined as a DFT[12] of the logarithm of the magnitude spectrum:

$$C(n) = DFT(\log(|X(k)|)) \qquad (7.54)$$

The significance of this is that we can work in the *cepstral* domain to process the
magnitude spectrum. If we remove the high-order cepstral coefficients, a process
called *liftering*, we can smooth the magnitudes. If we take these as if they were a

---

[12] In some places, the cepstrum is defined using the IDFT, although the forward transform was
used in the original paper that introduced the technique. Given that the magnitudes are real-valued,
it also makes sense to use the real DFT. In any case, such differences will not impact on the
application in question.

waveform, the fluctuations are equivalent to high-frequency components. Removing them smooths the 'wrinkles'. We can then recover the magnitudes by applying the inverse cepstrum:

$$|X(k)| = \exp(IDFT(C(n))) \qquad (7.55)$$

The amount of liftering determines how much smoothing is applied. Removing most coefficients will leave the overall shape, as in Fig. 7.14. Keeping a larger number of them will reveal individual partial peaks, if these are well separated. Depending on the application, we will have to decide how much liftering is necessary. In addition, to get a closer approximation, it is possible to use an iterative method known as the *true envelope* [91], which applies the cepstrum-liftering process repeatedly, comparing the result with the original magnitudes until a certain threshold is reached. In many applications, a single cepstrum is enough to give good a good approximation of the spectral envelope. Listing 7.16 shows a Python function implementing this process.

Listing 7.16 Spectral envelope extraction using cepstrum.

```python
import pylab as pl

def spec_env(frame,coefs):
    mags = abs(frame)
    N = len(mags)
    ceps = pl.rfft(pl.log(mags[:N-1]))
    ceps[coefs:] = 0
    mags[:N-1] = pl.exp(pl.irfft(ceps))
    return mags
```

With this, we can implement a spectral version of the vocoder instrument. The idea is that we extract the spectral envelope from one sound and apply it to an excitation signal. In the time-domain example, this is done via a bank of band-pass filters whose bands overlap, spaced evenly by a set frequency ratio. Here, the $N/2 + 1$ spectral bands play the role of the filter bank. Two inputs are used, one providing the phases and the other, the spectral envelope. These are combined to create a hybrid sound, to which the phase spectrum contributes the frequencies of the partials (the excitation part). The amplitude spectrum provides the filtering.

The code excerpt in Listing 7.17 illustrates the process. We extract the spectral envelope from both sounds. The ratio of the excitation magnitudes (`mags` in the code) and its spectral envelope (`env1`) flattens the amplitudes of the sound partials. To this we then apply the filter spectral envelope (`env2`) to shape their amplitude. This effect works very well if we use an excitation source rich in harmonics and a filter signal with a well delineated time-varying spectrum. A typical example uses an instrumental sound for the former and a voice for the latter. The full program for this process is presented in Appendix B (sect. B.8).

Listing 7.17 Spectral vocoder example.

```python
for n in range(0,L,H):
```

```
if(L-n < N): break
frame1 = stft(signal1[n:n+N],win,n)
frame2 = stft(signal2[n:n+N],win,n)
mags = abs(frame1)
env1 = spec_env(frame1,20)
env2 = spec_env(frame2,20)
phs = pl.angle(frame1)
if(min(env1) > 0):
  frame = p2r(mags*env2/env1,phs)
else:
  frame = p2r(mags*env2,phs)
output[n:n+N] += istft(frame,win,n)
```

7.2.3 Instantaneous frequencies

So far, we have only manipulated the magnitude spectrum, leaving the phases un-
touched. The reason for this is that it is very difficult to process them directly. The
phase spectrum, as we might have noticed, relates to the frequencies of partials, but
it is not yet clear how. To explore this, let's look again at a simple sinusoidal oscillator and the relationship between its phase and frequency. Recalling the discussion
in Chap. 3, Sect. 3.2.1, we see from Eq. 3.3 that to produce a $\sin(\phi(t))$ wave with
frequency f, we need a phase increment ϕ defined by

$$i_\phi(t) = \frac{2\pi d f(t)}{f_s} \tag{7.56}$$

and so we have $\phi(t+d) = \phi(t) + i_\phi(t)$, where $d = 1$.

This tells us of a fundamental relation between phase and frequency: to get the
next phase, we need to sum the current phase and a frequency-dependent increment,
that is, we *integrate* (with respect to time). Now, assuming we do not know the
frequency of an oscillator, but we have the phase values $\phi(t)$, how can we get the
corresponding frequencies $f(t)$ at each sample? Equation 7.56 can help us:

$$f(t) = \frac{f_s(\phi(t+d) - \phi(t))}{2\pi d} = \frac{f_s}{2\pi d}\Delta_\phi^d \tag{7.57}$$

The frequencies $f(t)$ are known as *instantaneous frequencies*, and we get them
by taking the phase difference $\Delta_\phi^d = \phi(t+d) - \phi(t)$, that is, we use the *derivative*
of the phase. So, these two operations, integration and differentiation, are used to
relate frequency to phase.

We are in a good position to estimate the instantaneous frequencies for each bin
of the STFT. We have the phases at each time point n, measured every l samples,
so we can use Eq. 7.57, with $d = l$. For one time point n, we use two consecutive
frames to get the phase difference:

$$\Delta^l_{\phi_n}(k) = \phi_{n+1}(k) - \phi_n(k) \tag{7.58}$$

The values of $\Delta^l_{\phi_n}(k)$ are, effectively, scaled and offset bin frequencies. To recover the phases, we just rearrange Eq. 7.58 to perform the integration:

$$\phi_{n+1}(k) = \phi_n(k) + \Delta^l_{\phi_n}(k) \tag{7.59}$$

For some applications, we can just use $\Delta^l_{\phi_n}(k)$ directly. For other uses, we might want to get the actual frequency in Hz that is detected for each bin, which has to take into account the scaling and offset that are implicit in this result. For this we use Eq. 7.57, which has the correct scaling factor. This gives the *deviation* from the bin centre frequency, which needs to be added in as an offset. The detected frequency $f(k)$ at time n is then

$$f_n(k) = k\frac{f_s}{N} + \frac{f_s}{2\pi l}\Delta^l_{\phi_n}(k) = f_c(k) + d_n(k) \tag{7.60}$$

Let's consider the meaning of this expression. If the detected frequency is exactly at the bin centre, $d_n(k) = 0$; $d_n(k) \neq 0$ makes the frequency deviate above or below this by a certain amount. So $f_n(k)$ is encoding a bin-by-bin FM process and, complementarily, $A_n(k)$, the amplitudes, encode an AM one. Another important point to note is that the range of values we should expect for $d_n(k)$ is determined by the analysis frame rate $f_r = \frac{f_s}{l}$:

$$|d_n(k)| \leq \frac{f_r}{2} \tag{7.61}$$

The minimum frame rate (and hop size l) is determined by the window bandwidth B_w, which will depend on its type and size. For the von Hann window, this is equivalent to

$$B_w = \frac{4f_s}{N_w} \tag{7.62}$$

where N_w is the window size (and also, in the cases examined here, the DFT size N). Therefore, the minimum range of $d_n(k)$ range for this window is $4\frac{f_s}{N_w}$ (ranging from $\frac{-2f_s}{N_w}$ to $\frac{2f_s}{N_w}$. For $N_w = 1024$ and $f_s = 44100$, this is ≈ 172 Hz, requiring a hop size l of 256 samples ($\frac{N_w}{4}$).

Since bin centre frequencies are spaced by f_s/N, this means that the von Hann window will also detect partials in adjacent bins on both sides of the centre, as shown in Fig. 7.15, where a partial at the centre of bin 16 is also picked up by bins 15 and 17 (with $N_w = N$). When the frequency is not at the centre, it will appear in another bin at the side (four bins in total if the frequency falls midway between two bins; see also Fig. 7.7).

With these principles in place, we are ready to convert the spectral data from a magnitude-phase representation to one in terms magnitudes and frequencies. This method is known as *phase vocoder* analysis [23, 41, 25, 71]. Since there will be

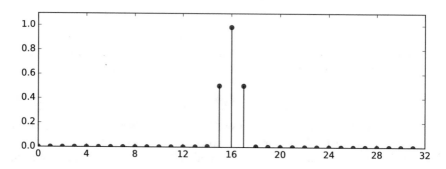

Fig. 7.15 Magnitude spectrum of $\sin(2\pi 16t/N)$ shaped by a von Hann window, with $N = 64$, where we can see that the analysis shows the partial in three bins, at the centre frequency and two adjacent points.

some state to be kept track of (the current phases), we implement this process as a Python class with two methods `analysis()` and `synthesis()` (Listing 7.18). We make use of the STFT functions defined earlier.

Listing 7.18 Phase vocoder class.

```python
import pylab as pl

def modpi(x):
    if x >= pl.pi: x -= 2*pl.pi
    if x < -pl.pi: x += 2*pl.pi
    return x
unwrap = pl.vectorize(modpi)

class Pvoc:
    def __init__(self,w,h,sr):
        N = len(w)
        self.win = w
        self.phs = pl.zeros(N//2+1)
        self.h = h
        self.n = 0
        self.fc = pl.arange(N//2+1)*sr/N
        self.sr = sr

    def analysis(self,x):
        X = stft(x,self.win,self.n)
        delta = pl.angle(X) - self.phs
        self.phs = pl.angle(X)
        delta = unwrap(delta)
        freqs = self.fc + delta*self.sr/(2*pl.pi*self.h)
```

```
        self.n += self.h
        return abs(X), freqs

    def synthesis(self,a,f):
        delta = (f - self.fc)*(2*pl.pi*self.h)/self.sr
        self.phs += delta
        X = a*pl.cos(self.phs) + 1j*a*pl.sin(self.phs)
        y = istft(X,self.win,self.n)
        self.n += self.h
        return y
```

The processes are implemented as outlined above. The only extra detail is that we need to keep the phase deltas within the principal-value range ($-\pi$ to π), otherwise the frequency calculation might fail to give us the correct values in Hz. For this, we write a simple modulus function and use `pl.vectorize` to allow it to iterate directly over a NumPy array. This may be ignored if we are keeping just the phase differences. The test program below can be used to verify that the process works. It also prints the frequency found in the bin where the sine wave should have been detected.

Listing 7.19 Phase vocoder test program.

```
import pylab as pl

N = 1024
D = 4
H = N//D
win = pl.hanning(N)
sr = 44100
f = 50
t = pl.arange(0,sr)
L = sr
t = pl.arange(0,L)
signal = pl.sin(2*pl.pi*f*t/sr)
output = pl.zeros(L)
scal = 1.5*D/4
pva = Pvoc(win,H,sr)
pvs = Pvoc(win,H,sr)
cf = round(N*f/sr)


for n in range(0,L,H):
    if(L-n < N): break
    amps,freqs = pva.analysis(signal[n:n+N])
    print(round(freqs[cf]))
    output[n:n+N] += pvs.synthesis(amps,freqs)

pl.figure(figsize=(8,3))
```

```
pl.ylim(-1.1,1.1)
pl.plot((output/scal),'k-')
pl.show()
```

Now we can manipulate the frequencies as well as the amplitudes. One application is to build a pitch shifter that can take in a transposition ratio and scale all frequencies according to it. To do this, we have to move the data from one bin to another, because as the components get scaled, they end up around a different centre frequency. If we did not shift the data, the frequencies would exceed the bandwidth of a given bin and the scaling would not work. So, for a pitch shift factor p, we apply the following mapping for amplitudes $A(k)$ and frequencies $f(k)$:

$$A_{out}(pk) = A_{in}(k) \qquad 0 < kp < N/2 - 1 \tag{7.63}$$

$$f_{out}(pk) = p f_{in}(k) \qquad 0 < kp < N/2 - 1 \tag{7.64}$$

This is implemented in the Python function `scale()` in Listing 7.20. This code excerpt uses the previously defined `pv` class to perform the analysis and synthesis.

Listing 7.20 Spectral pitch shifter.

```
import pylab as pl

def scale(amps,freqs,p):
    N = len(freqs)
    a,f = pl.zeros(N),pl.zeros(N)
    for i in range(0, N):
        n = int(round(i*p))
        if n > 0 and n < N-1:
            f[n] = p*freqs[i]
            a[n] = amps[i]
    return a,f

for n in range(0,L,H):
    if(L-n < N): break
    amps,freqs = pva.analysis(signal[n:n+N])
    amps,freqs = scale(amps,freqs,trans)
    output[n:n+N] += pvs.synthesis(amps,freqs)
```

The quality of the effect will be enhanced if we have a shorter hop size l. A value of $N/8$ should be considered as the minimum for this application (using the von Hann window). This process will work well for many different types of input, with the added benefit that it has built-in aliasing suppression, since we cannot scale beyond the last bin. For some sounds, such as vocals, formant preservation is key and so we need to bear that in mind. The present code scales all formants, so it does not work well for voices, unless a *Donald Duck* effect is desired.

We already have a solution for this, as we can extract the original spectral envelope and apply it to the shifted sound. For this particular application, it might be worth implementing the true envelope algorithm [91], as outlined in Sect. 7.2.2. This is shown in listing 7.21, where the `true_spec_env()` function takes in amplitudes, calculates their logarithm and then enters into an iterative process. This applies the cepstrum and liftering, and then recovers the log magnitudes, until the difference between these and the cepstrum input data is equal to or below a certain threshold. As can be seen, this process involves more computation than the straight cepstrum operation, and thus it is not as efficient (although it can produce superior results).

Listing 7.21 Formant preservation.

```python
import pylab as pl

def true_spec_env(amps,coefs,thresh):
    N = len(amps)
    sm = 10e-15
    form = pl.zeros(N)
    lmags = pl.log(amps[:N-1]+sm)
    mags = lmags
    check = True
    while(check):
        ceps = pl.rfft(lmags)
        ceps[coefs:] = 0
        form[:N-1] = pl.irfft(ceps)
        for i in range(0,N-1):
            if lmags[i] < form[i]: lmags[i] = form[i]
            diff = mags[i] - form[i]
            if diff > thresh: check = True
            else: check = False
    return pl.exp(form)+sm

for n in range(0,L,H):
    if(L-n < N): break
    amps,freqs = pva.analysis(signal[n:n+N])
    env1 = true_spec_env(amps,40,0.23)
    amps,freqs = scale(amps,freqs,trans)
    env2 = true_spec_env(amps,40,0.23)
    output[n:n+N] += pvs.synthesis(amps*env1/env2,freqs)
```

As before, we extract both pre- and post-transposition envelopes and scale the amplitudes by their ratio, so that the original spectral envelope gets applied to the flattened magnitudes of the scaled spectrum. Appendix B (B.9) provides the full program. Other processes involving frequency manipulation include shifting (rather than scaling) by an offset, interpolating two input spectra to implement morphing, time smoothing of frequency data for a blurring effect, and spectral delays.

7.2.4 Time scaling

The ideas about phase vocoder developed in the previous section allow us to treat each frame separately and move from one to another at different rates, including stopping (freezing) at one of the time points. This makes phase vocoder encoding a very flexible way to describe a sound as a collection of short snippets of sound kept in spectral form. In some ways, we can think of each frame as a grain, in the way discussed in Chap. 6. However, this time, we can obtain a cleaner result when stretching, compressing, freezing or jumbling up the original time scale of sounds.

To implement various time scale modifications, we can simplify the STFT and instantaneous frequency analysis. We do not need to convert phase differences into frequencies in Hz, so we can skip some steps in the process. For instance, the STFT becomes simply[13]

```
def stft(x,w):
    X = pl.rfft(w*x)
    return X
```

as we do not need to worry about any phase shifts (they will cancel out in the integration). The ISTFT can be similarly minimal.

Here's the outline of the time-scaling process:

1. Take the audio in and fill two frames, separated by one hop size.
2. Take the STFT and the inter-frame phase differences (frequencies).
3. Calculate the current phases by the adding the previously-output phases to the estimated frequencies.
4. Make a frame composed of the magnitudes of one of the two analysis frames and the current phases.
5. Take the ISTFT of the output frame and overlap-add it at a constant hop size.

This makes it possible for us to jump to any position in the input audio source and nearly seamlessly produce a smooth output signal. In other words, we de-couple the analysis and synthesis time positions. For this to work well, it is important that the hop size of the two operations are matching. This is the reason for taking two input STFT frames at a time (since we cannot guarantee that the next analysis point will be a fixed hop size ahead of the current time).

A simple demonstration is provided by a time-scaling program that stretches or compresses the duration of a sound according to a given ratio (without affecting pitch). This is shown in the code excerpt presented in Listing 7.22.

Listing 7.22 Time-scaling example.

```
import pylab as pl


for n in range(0,L-(N+H),ti):
```

_____

[13] In fact this form can be used for all amplitude-only manipulations such as masking and cross-synthesis.

```
X1 = stft(signal[n:n+N],win)
X2 = stft(signal[n+H:n+H+N],win)
diffs = pl.angle(X2) - pl.angle(X1)
phs += diffs
Y = p2r(abs(X2),phs)
output[np:np+N] += istft(Y,win)
np += H
```

In this case, the hop size is H, the DFT size is N, and the functions stft(), stft(), and p2r() are as given before. The time increment ti is the product of the time scale ratio and the hop size. If the ratio is above one, the increment is more than one hop size and we have compression (time will proceed faster than in the original). Conversely, if the ratio is below one, then the time will be stretched.

One artefact that is often reported in time-stretching applications is a certain *phasiness* that results from the process. This is more pronounced for extreme time scale ratios. The main cause of this is that the re-constructed bin phases get out of sync, they become misaligned with regards to the original spectrum. This introduces beating effects between adjacent bins, due to the fact that a partial is often detected by more than one bin at the same time, as it falls within the bandwidth. In the case of the von Hann window, as we have seen in section 7.2.3, a component might be found in two or three extra bins adjacent to the one whose centre is closest to its frequency[14].

The solution for this is to try to lock the resynthesis phase to the value at the centre (determined by a magnitude peak). This can be done by peak-picking, but a more practical and effective solution is to apply a three-point averaging across the output phases encoded as a complex number [85]:

$$e^{\phi_{locked}(k)} = e^{j\phi(k)} - \left(e^{j\phi(k-1)} + e^{j\phi(k+1)}\right) \tag{7.65}$$

where $\phi(k)$ is the output phase of bin k. This averaging uses different signs for adjacent bins, as the components are likely to be half a period (π radians) out of phase in the original analysis.

To apply phase locking, we can create a function that takes in phases, put them into complex form (using the p2r function with unit magnitudes) and perform the averaging, obtaining the locked phases as a result. This is shown in Listing 7.23.

Listing 7.23 Time scaling with phase locking.

```
import pylab as pl


def plock(phs):
    N = len(phs)
    x = p2r(pl.ones(N),phs)
    y = pl.zeros(N)+0j
    y[0], y[N-1] = x[0], x[N-1]
```

---

[14] It is possible to check this out by printing two extra bins, one at each side of the centre, in the phase vocoder test program.

```
    for i in range(1,N-1):
        y[i] = x[i] - (x[i-1] + x[i+1])
    return y

for n in range(0,L-(N+H),ti):
    X1 = stft(signal[n:n+N],win)
    X2 = stft(signal[n+H:n+H+N],win)
    diffs = pl.angle(X2) - pl.angle(X1)
    phs += diffs
    Y = p2r(abs(X2),phs)
    output[np:np+N] += istft(Y,win)
    phs = pl.angle(plock(phs))
    np += H
```

However, at this point we appear to have been going back and forth between rectangular and polar representations. As pointed out in [85], for this particular algorithm, it is possible to avoid such conversions, keeping the whole computation in rectangular form. Let's consider first the original form, without phase locking. The output spectrum at time $t + r$, for an arbitrary r, can be written in polar form (with a complex exponential) as

$$Y_{t+r}(k) = A_{t+l}(k)e^{j(\phi_{t+l}(k)-\phi_t(k)+\phi_{t+r-l}(k))} \tag{7.66}$$

where l is the hop size and $X_t(k) = A_t(k)e^{j\phi_t(k)}$. We can now use the following identities:

$$A_{t+l}(k)e^{j(\phi_{t+l}(k)-\phi_t(k))} = |X_{t+l}(k)|\frac{X_{t+l}(k)}{X_t(k)}\frac{|X_t(k)|}{|X_{t+l}(k)|} = X_{t+l}(k)\frac{|X_t(k)|}{X_t(k)} \tag{7.67}$$

$$e^{j\phi_{t+r-l}(k)} = \frac{Y_{t+r-l}(k)}{|Y_{t+r-l}(k)|} \tag{7.68}$$

Combining 7.67, 7.68 and 7.66, we have

$$Y_{t+r}(k) = X_{t+l}(k)\frac{Y_{t+r-l}(k)}{X_t(k)}\frac{|X_t(k)|}{|Y_{t+r-l}(k)|} \tag{7.69}$$

Thus we can simplify the time scaling implementation to:

Listing 7.24 Time scaling in rectangular form.

```
import pylab as pl

Y = pl.zeros(N//2+1)+0j+10e-20
for n in range(0,L-(N+H),ti):
    X1 = stft(signal[n:n+N],win)
    X2 = stft(signal[n+H:n+H+N],win)
    Y = X2*(Y/X1)*abs(X1/Y)
```

```
output[np:np+N] += istft(Y,win)
np += H
```

Note that we have had to make sure that the initial value of Y is not zero, otherwise the recursion would not work (it would always produce zeros). Care might be necessary to avoid divisions by zero as well. To complement this, we can introduce the phase-locking operation, this time without any need for conversions. To make the code more compact and efficient, we pack the loop into three array operations.

Listing 7.25 Time scaling in rectangular form, with phase locking.

```
def plock(x):
    N = len(x)
    y = pl.zeros(N)+0j
    y[0], y[N-1] = x[0], x[N-1]
    y[1:N-1] = x[1:N-1] - (x[0:N-2] + x[2:N])
    return y

Z = pl.zeros(N//2+1)+0j+10e-20
for n in range(0,L-(N+H),ti):
    X1 = stft(signal[n:n+N],win)
    X2 = stft(signal[n+H:n+H+N],win)
    Y = X2*(Z/X1)*abs(X1/Z)
    Z = plock(Y)
    output[np:np+N] += istft(Y,win)
    np += H
```

These principles can also be applied to other spectral processes that do not require the analysis to be expressed in terms of separate amplitudes and frequencies, or even to processes that only manipulate amplitudes. In such cases, this approach might turn out to be more convenient and/or efficient. A full program implementing the time scaling process outlined in this section can be found in Appendix B (Sect. B.10).

7.3 Real-Time Spectral Processing

So far in this study, we have not touched on the question of real-time processing, as we were concentrating on examining the techniques without reference to where we would apply them. The examples in Python have taken their inputs from an array in memory and placed them into another array. Although the full programs in Appendix B use soundfiles as the source and destination of the data in memory, that is only for convenience. We could instead have used the input arrays as buffers into which data from a soundcard was transferred, and similarly taken from the output array into the soundcard. Even though Python is not optimised for speed and the code did not attempt to be particularly efficient, it is likely that on a modern computer all example programs could run with real-time output (and input whenever appropriate).

However, we can do much better by using Csound, which provides excellent support for real-time spectral processing [55, 59]. It has a suite of opcodes that implement *streaming* analysis, modification and synthesis. For this, these opcodes use a specially-designed data type, *fsig*, that holds a time-ordered sequence of spectral frames (instead of audio samples). These opcodes can be mixed seamlessly with other time-domain processing. An fsig can be treated very much like any other type of signal and passed around from one opcode to another.

We can generate fsigs from a streaming analysis using `pvsanal`, which takes time-domain audio input and produces a sequence of frames at every time point (separated by hop size samples). The parameters of this opcode are the usual DFT size, hop size, window size and type. The window can be larger than the DFT size, which in some cases might produce a cleaner analysis, but in general we can set it to the same value as the DFT size. Non power-of-two lengths are allowed, but are more computationally demanding. The resynthesis is performed by `pvsynth`, which only takes an fsig as a parameter. Other sources for fsigs include file and table readers and we can also resynthesise them with a bank of oscillators (additive synthesis).

Csound implements a large number of different processes that can be applied to fsigs, including all of the ones explored in this chapter. As an example, the following instrument implements real-time pitch shifting with formant preservation as in the example discussed above, creating an extra harmonising voice above the original.

Listing 7.26 Harmoniser instrument using streaming spectral processsing.

```
instr 1
  asig = in()
  iN = 1024
  iH = iN/8
  fs1 = pvsanal(asig,iN,iH,iN,1)
  ahrm = pvsynth(pvscale(fs1,1.25,1))
  amix = ahrm+delay(asig,(iN+iH)/sr)
    out(amix/3)
endin
```

In this example, we take the input signal and create an fsig, which is then transposed by 1.25 (a major third), keeping formants, and resynthesised. The original signal needs to be delayed by one frame and one hop size, to align it correctly with the processed output. The reason for this is that the streaming process needs to use two STFTs one hop size apart to calculate the phase vocoder data (as we have seen in Sect. 7.2.4. For this we need to wait for enough data to come in from the input, which is exactly DFT size + hop size samples. So spectral processing inherently introduces some latency between input and output if we are streaming data from a source in real time. No such problem exists if we are reading from memory or from a file. A full discussion of spectral processing in Csound can be found in [59].

7.4 Conclusions

Spectral processing is a significant area in DSP research and provides a great variety of very interesting effects for sound design, music performance and composition. In this chapter, we were able to introduce some key concepts of frequency-domain methods. The code examples, in the form of excerpts or full programs, provided an illustration of what is possible with these tools. Readers should be able to expand on these to create their own customised manipulations of spectral data.

We have noted that spectral processing can be applied in real time in the development of computer instruments. There is fertile ground here for innovation, and it is possible to play with these processes in an interactive and experimental way. Such an approach is encouraged, as many interesting new sounds can be discovered through the tweaking of algorithms and parameters.

Finally, we should note that spectral analysis can be paired very successfully with additive synthesis. In the case of the amplitude-frequency data format of the phase vocoder, we can resynthesise directly from a spectral frame using a bank of oscillators. This might not be as efficient as the inverse STFT operation, but it allows more flexibility, since the oscillator frequencies are not bound to strict bin bandwidths and we can also choose to resynthesise a reduced set of bins. Further processing can be applied to frame data to generate partial tracks that can be used as input to oscillators. In this case, we need to find spectral peaks in each frame and then link these in time to form a sequence. There are a variety of possibilities that are open to this type of approach. More details of such techniques can be found in [6, 66, 59, 80, 93, 97].

Part III
Application Development

Chapter 8
Interaction

Abstract This chapter examines the wide-ranging area of interaction with computer instruments from two perspectives: communications and interfaces. The former is presented as the binding element that connects together the parts of a complex system. The latter relates to the kinds of user experience that are possible for the types of software we have been exploring in this book. In the course of the text, we will introduce the key aspects of MIDI and OSC, and examine the possibilities for the provision of graphical and physical interfaces to computer instruments.

Interaction with computer instruments can take many forms. So far, we have limited the range of interfaces employed to various types of text input, through which we were able to design and build the signal processing components, as well as construct various types of parameter automation and control structures. In this chapter, we will examine how we can extend the level of interaction to other types of interfaces, which can take the form of software or hardware objects. As part of this, we will study protocols for communication that allow us to use interfaces that are external to the computer and can be extended to contain remotely-located components. This will be complemented by a review of the various types of graphical and custom hardware options for the design of interaction with computer instruments.

8.1 Communication Protocols

Electronic music instruments have benefitted from the development of standard protocols for the exchange of control data. Prior to their development, only ad-hoc and hardware-specific types of connections were possible. With the advent of the Musical Instrument Digital Interface (MIDI), the development of network communications, Open Sound Control (OSC), and the widespread use of standard serial interfaces, many possibilities now exist for the design of instruments that can interact with a variety of software and hardware controllers, both locally and remotely placed.

© Springer International Publishing AG 2017
V. Lazzarini, *Computer Music Instruments*,
https://doi.org/10.1007/978-3-319-63504-0_8

8.1.1 The MIDI protocol

MIDI is a communication protocol that was originally designed to connect electronic music instruments such as synthesisers, drum machines and sequencers [67]. Over the years, it became implemented in a wider range of devices, including computers, as the popularity of microprocessor-based systems grew. Today, it is still a very widespread technology for connecting different types of instruments, although its limitations can pose difficulties when increased precision and/or flexibility is required.

On the hardware side, the MIDI protocol originally dictated that device links used a 5-pin DIN connector, with optically-isolated connectors to avoid ground loop interference in the audio signal, running at a rate of 31,250 baud[1]. Since 1999, a specification for the sending of MIDI over a Universal Serial Bus (USB) connection has been in place, allowing instruments to communicate with a host at higher speeds, although the actual implementation details can vary from one device to another. MIDI connections through 5-pin DIN are unidirectional, carrying the data from a MIDI-out to a MIDI-in device. USB connections are host-oriented, connecting a MIDI instrument to a host such as a computer, but the data can be carried from host to device as well as from device to host, depending on the hardware in question.

A MIDI connection is defined to happen at a port (IN, OUT), which contains 16 logical channels. Ports in a computer can be physical (e.g. hardware-based, connected to external devices), or virtual (software) allowing us to send messages from one program to another. These connections carry a MIDI message, which is composed of a *status* byte optionally followed by one or two data bytes. Messages are of two types: system and channel. The latter applies to a specific channel, whose number is indicated in the status byte, while the former does not make such a distinction and does not carry any information regarding channels.

Channel messages

Channel messages can be of either a voice or a mode sub type. In order to send or receive these from a language such as Python, we can use a MIDI library or module that allows us to access physical or virtual ports (of which there might be more than one available in a system). In this chapter, examples will be given using `mido`. The following code,

```
import mido
o = mido.open_output('port name')
```

imports the library and opens a port called `'port name'`, which should exist in the system. The function can be called with no parameters to select the default output (set by the system). The variable o holds the open port and can be used to send output messages to it.

---

[1] The Baud is a unit of data transfer rate usually defined as *symbols per second*. In the case of MIDI, since each symbol is 1 bit, it is equivalent to bits per second.

On the software synthesis side, generally a complete MIDI implementation will be provided. In the case of Csound, we can get input from a port via the -M <port_name> option. This can be either a software or a hardware port. Faust stand-alone programs will create a new software port in the system carrying the name the program was compiled to.

Channel voice

Channel voice messages are usually employed to send commands to synthesisers to start or stop a sound, shape controls, modify pitch, or change presets. Each message has a numeric code to identify it, together with a channel. Values generally range from 0 to 127. There are seven types of these messages:

1. **NOTE ON**: code 144, contains two data values. The first is a note number, and the second a note velocity (amplitude):

```
o.send(mido.Message('note_on', note=60,
        velocity=64,channel=1))
```

Note numbers range from 0 to 127. They can be mapped to any frequency, but usually this is done to a 12TET scale where A4 is note number 69.

2. **NOTE OFF**: code 128, contains two values, a note number and velocity. This message can also be emulated with a NOTE ON containing zero velocity:

```
o.send(mido.Message('note_off', note=60,
        velocity=64,channel=1))
```

3. **Aftertouch**: code 208, contains a single value, the amount of aftertouch. This is typically generated by applying pressure to the controller keyboard:

```
o.send(mido.Message('aftertouch', value=100,
        channel=1))
```

4. **Polyphonic aftertouch**: code 166, carries a note-by-note aftertouch control with two data values (note number and amount):

```
o.send(mido.Message('polytouch',note=64,
        value=100,channel=1))
```

5. **Program change**: code 192, contains a single value, the program number. This is often used to change stored parameter programs in synthesisers:

```
o.send(mido.Message('program_change', program=100,
        channel=1))
```

6. **Control change**: code 176, contains two values, a control number and a value. This can normally be mapped by the user. Some controller numbers are set by default (e.g. 1 is modulation wheel in keyboard controllers):

```
o.send(mido.Message('control_change', control=10,
        value=90,channel=1))
```

7. **Pitch bend**: code 224, carries two values that can be combined to give a wider range than the other messages, from −8192 to 8191.

```
o.send(mido.Message('pitchwheel',pitch=8191,
        channel=1))
```

Csound can respond to these messages with specific opcodes. Each NOTE_ON and NOTE_OFF will start and stop an instrument instance, which by default is assigned to its channel number (`instr` 1 to channel 1, etc.). This mapping can be changed. A simple example of a polyphonic synthesiser with controls for amplitude (from NOTE_ON velocity and polyphonic aftertouch), pitch (from NOTE_ON note number plus pitch bend control) and filter cut-off frequency (from MIDI control 1, normally assigned to the modulation wheel) is shown in Listing 8.1. This design allows a performer to change timbre through a modulation control, and to apply AM (tremolo) through aftertouch, and FM (bend/vibrato) via a pitch wheel.

Listing 8.1 Csound MIDI synth example

```
instr 1
 kcf init 1
 ia = ampmidi(0dbfs/8)
 kf = cpsmidib:k(2)
 ka = polyaft:k(notnum(),0.5,1.5)
 asig = vco2(ia*ka,kf)
 kcf  = midic7:k(1,1,8)
 afil = moogladder(asig,kcf*kf,0.9)
    out(linenr:a(afil,0.01,0.1,0.01))
endin
```

As mentioned above, Faust stand-alone programs can also make use of MIDI message inputs. These are mapped directly to UI controls (sliders, buttons, etc.) by including the following codes as part of the the UI name:

- `[midi:ctrl n]`: control change, n is the controller name.
- `[midi:keyon n]`, `[midi:keyoff n]`: note velocity, n is the MIDI note number (NOTE_ON, NOTE_OFF messages).
- `[midi:keypress n]`: polyphonic aftertouch, n is note number.
- `[midi:pitchwheel]`: pitch bend values (0-16384 range).

Controls are mapped to the (0,127) range for continuous UI controls (e.g. sliders), or to 0/127 for switches (e.g. buttons). Additionally, polyphonic capabilities can also be built into a compiled program. This involves a maximum number of Faust DSP instances that can be run at the same time, which can be defined at compile or run time. In this case, the program will also respond to the following UI control names:

- `gate`: note gate, 1 when one of the DSP voices is on, 0 when off.
- `freq`: note frequency, mapped into 12TET with A4 = 440 Hz set to note number 69.

- `gain`: gain, from note velocity, mapped between 0 and 1.

As an example, the FBAM2 instrument (see Sect. 5.4.1) can be turned into a MIDI-capable instrument by setting up some MIDI UI controls for it:

Listing 8.2 Faust MIDI synth example

```
gate = button ("gate") : lp(10);
amp = hslider("gain", 0.1, 0, 1, 0.001)*gate;
freq = hslider("freq[unit:Hz]",440,27.5,4435,1);
beta = hslider("beta[midi:ctrl 1]",90,0,127,1)
        : *(1./127) : lp(3);

lp(freq) = *(1 - a1) : + ~ *(a1) with
    { C = 2. - cos(2.*PI*freq/SR);
      a1 = C - sqrt(C*C - 1.); };

rms = fabs : lp(10);
balance(sig, comp) =  sig*rms(comp)/(rms(sig)+0.000001);
fbam2(beta) = ((+(1))*(_)) ~ (_ <: _ + _')*(beta));
process = amp*(osci(freq) <: fbam2(beta),_ : balance);
```

The code in Listing 8.2, if compiled with polyphonic and MIDI support to, for instance, an application called *fbamidi* will respond to MIDI commands sent to a port with that name[2].

Channel mode

Channel Mode messages use the same code as a control change, with the first data byte set between 120 and 127. These messages define how an instrument should receive any voice message that follows. Mode messages are sent on the *basic* channel to which a receiver is assigned. The number of a basic channel cannot be changed via a channel mode or voice message. The following are the eight channel mode message types:

1. **All sound off** (data byte = 120): turns all sound off in the receiver (including effects, etc).
2. **Reset all controllers** (121): resets controller configurations/conditions to initial state.
3. **Local control** (122): disconnects the local controller from the local synthesis engine (0, local off) or reconnects it (1, local on).
4. **All notes off** (123): requests the receiver to send NOTE_OFF messages to all of its channels.
5. **Omni off** (124): requests the receiver to recognise the channel of voice messages.
6. **Omni on** (125): requests the receiver to ignore the channel of voice messages.

---

[2] The online Faust compiler at http://faust.grame.fr/onlinecompiler can be used for this purpose.

7. **Mono on** (126): requests the receiver to respond monophonically to note messages.
8. **Mono off** (127): requests the receiver to respond polyphonically to note messages.

System messages

System messages are of three subtypes: exclusive, common, and real time:

1. **System exclusive**: these messages are used to send device-specific data, such as parameter lists, sample dump, etc.
2. **System common**: these are used to send MIDI time code synchronisation messages, song position and selection, and oscillator tuning request.
3. **System real time**: all messages used to control the MIDI clock and provide uniform timing information for systems that rely on MIDI for synchronisation.

8.1.2 The OSC protocol

OSC was developed as an alternative to the MIDI protocol [108, 109] and has been adopted widely by the computer music community. It employs internet protocol (IP) networks as a medium for message delivery, through the internet internet transport layer protocol (UDP). Its applications are not limited to typical electronic music instruments, as it is designed to support various types of computer music interaction.

An OSC message is composed of the following elements:

1. Destination IP address.
2. Destination port.
3. Address.
4. Type.
5. Data.

The message is sent to a receiver with a given IP address, generally of the IP version 4 form composed of four 8-bit bytes (usually formatted as four numbers separated by dots). The address can also use lookup, in which case it can be a name (e.g. *name.domain*). The local machine can be identified as 'localhost' or as '127.0.0.1'. Following this, a network port is required and this needs to be set by the receiving end (generally, a four-digit or greater number). This can be any arbitrary free port on the machine.

OSC mandates that a message has an *address* that is a string of the form */destination[/subdestination/...]*. This is defined by the receiving end (the host). Any number of addresses can be defined and these are not tied to any predefined names. An address needs to be accompanied by a message *type*, which is specific to it. Messages are only received if the address and the type match. The type is also a string defining the data carried by the message. For each data object, one of the following type letters needs to be added:

- `"i"`: 32-bit integer.
- `"f"`: float (single-precision).
- `"s"`: string.
- `"b"`: blob (unspecified arbitrary-sized data).

In addition to these standard types defined by version 1.0 of the specification, hosts are allowed to define other types of data. If undefined, messages carrying them are discarded. Some examples are:

- `"h"`: 64-bit integer.
- `"d"`: double-precision float.
- `"c"`: single character.
- `"[.]"`: array (with a data type in place of '.').

Both Python and Csound, as well as stand-alone Faust programs, can avail of OSC support. In Csound, we have the opcodes `OSCinit` and `OSClisten`, for setting up a server and listening to messages, and `OSCsend` for sending them. For Python, there is a `pythonosc` module that supports server and host functionalities.

OSC works with the concept of server-client communications. A server is set up to respond to any number of OSC message addresses on a given machine and port. Any client can then send messages to this IP address and port to communicate with the server. As an example, the following program will establish a server using Csound and send OSC commands to it from Python.

The server is organised thus:

1. A source-filter synthesis instrument (similar to the one in Listing 8.1) is set up, with amplitude, frequency and filter cut-off controls.
2. An OSC listener instrument is designed to respond to on/off messages to start and stop instances of the synthesis instrument.
3. Global pitch can also be modified through OSC commands.
4. The listener responds to instance-oriented amplitude and cut-off frequency messages.

The code is shown in Listing 8.3. Instance on/off messages have the addresses "/inst/on" and "/inst/off" respectively. The former carries three data objects: instance number, frequency and amplitude, whereas the latter only needs an instance number. To start an instrument for an indefinite duration, we set p3 to -1 when scheduling an event; it can be stopped with a negative p1. The instrument numbers will have a fractional part that identifies a given instance. The code is set up for up to 100 distinct instances. To carry the per-instance amplitude and cut-off frequency data, we use global arrays. The value of pitch bend, given that it affects all instances, is carried in a single global scalar variable.

Listing 8.3 Csound OSC synth server example

```
giH OSCinit 40000
gka[] init 100
gkf[] init 100
gkp init 1

instr 1
 k1,k2,k3 init 0
 ka,kf init 0
 ks = 1
 while ks != 0 do
  ks OSClisten giH,"/inst/on","iff",k1,k2,k3
  schedkwhen(ks,0,100,2+k1/100,0,-1,k3,k2)
 od
 ks = 1
 while ks != 0 do
  ks OSClisten giH,"/inst/off","i",k1
  schedkwhen(ks,0,100,-(2+k1/100),0,0,0,0)
 od
 ks = 1
 while ks != 0 do
  ks OSClisten giH,"/filter/cf","if",kf,k2
  gkf[kf] = ks > 0 ? k2 : gkf[kf]
 od
 ks = 1
 while ks != 0 do
  ks OSClisten giH,"/amp","if",ka,k2
  gka[ka] = ks > 0 ? k2 : gka[ka]
 od
 ks = 1
 while ks != 0 do
  ks OSClisten giH,"/pitch","f",k1
  gkp = ks > 0 ? k1 : gkp
 od
endin
schedule(1,0,-1)

instr 2
 kcf init 1
 ia = p4
 kf = p5*gkp
 ka = gka[frac(p1)*100]
 asig = vco2(ia*ka,kf)
 kcf  = gkf[frac(p1)*100] + 1
 afil = moogladder(asig,kcf*(1+kf),0.7)
```

```
      out(linenr:a(afil,0.01,0.1,0.01))
endin
```

On the client side, we need to prepare messages to send to this server. With Python-OSC, we define the destination, and then package and send the messages to it. The code in Listing 8.4 does these two actions, starting four instances with different pitches, changing parameters through OSC commands.

Listing 8.4 Python OSC client example, controlling the server in Listing 8.3.

```python
from pythonosc import osc_message_builder as mb
from pythonosc import udp_client as ud
import time

dest = ud.UDPClient("127.0.0.1",40000)

frs = [220., 130., 390., 660.]

def pack_and_send(dest,m):
  msg = mb.OscMessageBuilder(address=m[0])
  for i in m[1:]: msg.add_arg(i)
  dest.send(msg.build())
  print('Message sent: %r' % m)

for i in  range(1,5):
    pack_and_send(dest,['/amp',i,1.])
    pack_and_send(dest,['/filter/cf',i,1.])

for i in  range(1,5):
  pack_and_send(dest,['/inst/on',i,frs[i-1],5000.])
  time.sleep(1)

for i in  range(1,5):
  pack_and_send(dest,['/filter/cf',i,3.0])
  time.sleep(1)

time.sleep(1)
pack_and_send(dest,['/pitch',4/3])
time.sleep(2)
pack_and_send(dest,['/pitch',1.])
time.sleep(1)

for i in  range(1,5):
  pack_and_send(dest,['/filter/cf',i,0.5])
  time.sleep(1)

for i in  range(1,5):
```

```
pack_and_send(dest,['/inst/off',i])
time.sleep(1)
```

Similarly to MIDI, Faust standalone programs can be built with OSC support that maps automatically any UI element to an OSC message address. If we take, for instance, the following Faust code (Listing 8.5), and compile it as a program called *nblsum* with OSC support, we will have a server running (by default) on port 5510.

Listing 8.5 Faust synthesiser program

```
vol = hslider("amp", 0.1, 0, 1, 0.001);
freq = hslider("freq",440,110,1760,1);
a = hslider("a", 1,0,1,0.001);

pi = 3.141592653589793;
sr = 44100;
mod1(a) = a - floor(a);
incr(freq) =  freq / float(sr);
phasor(freq) =  incr(freq) : (+ : mod1) ~ _ ;
w = 2*pi*phasor(freq);
th = 2*pi*phasor(freq*2);
sig = a*sin(w - th)/(1. - 2*a*cos(th) + a*a);
process = vol*sig;
```

This server will respond to control messages ('float' type) on the following addresses: /nblsum/amp, /nblsum/freq, and /nblsum/a. The following shows an example of a client for this code:

Listing 8.6 Python OSC client example, controlling the server in Listing 8.5.

```
from pythonosc import osc_message_builder as mb
from pythonosc import udp_client as ud
import time

dest = ud.UDPClient("127.0.0.1",5510)

pars = [(220.,0.1,0.8),
        (261.,0.2,0.6),
        (330.,0.4,0.4),
        (440.,0.8,0.3)]

def pack_and_send(dest,m):
  msg = mb.OscMessageBuilder(address=m[0])
  for i in m[1:]: msg.add_arg(i)
  dest.send(msg.build())
  print('Message sent: %r' % m)

for i in pars:
    pack_and_send(dest,['/nblsum/freq',i[0]])
```

```
pack_and_send(dest,['/nblsum/amp',i[1]])
pack_and_send(dest,['/nblsum/a',i[2]])
time.sleep(1.5)
```

OSC is extremely flexible. Together with the possibility of connecting to other machines over a network, it provides a very useful means for controlling computer music instruments.

8.1.3 Networks

While OSC is a convenient way to format, send and receive messages using an IP network, there is nothing to stop computer instruments using this infrastructure directly. This would of course require that the messages are understood on both sides of the connection and so this would not be universally applicable, although it might make sense in some situations. One advantage is that, if we have a specific type of data to be passed, there is no need to use a larger framework designed for various unspecified message formats. One such case is the server mode in Csound, which simply listens to complete text strings containing code and compiles them as soon as they are received. Any program that can send text strings through a network connection can use this facility, without any extra formatting requirements.

Network connections over IP are of two basic transport layer types: UDP, which has already been introduced as it is used by OSC; and transmission control protocol, TCP. There are significant differences between the two. UDP has the following characteristics:

- Independent, separate, unordered, data packets (datagrams).
- Transaction-based.
- No guarantees of message delivery, unreliable.
- Lightweight.
- Suitable for broadcast (many receivers).

TCP has a completely different set of characteristics:

- Data streams, guaranteed delivery (data is re-transmitted if lost).
- Connection-based.
- Reliability handling and congestion control.
- Heavyweight.
- Client-server connections.

It is clear that these two protocols have distinct applications. TCP connections require a structured link between two machines, whereas UDP is simply based on sending data to a destination (as we have seen in the case of OSC, clients know about a server, but a server knows nothing about its clients). TCP can be used, for instance, in a fairly stable network with a guaranteed quality-of-service (QoS) level to send audio data from one machine to another. UDP can also be used for the delivery of audio through an IP network, via the real-time transport protocol (RTP) which is built on top of it. In this case, UDP is favoured over TCP because it places timeliness ahead of reliability. The need for messages to arrive on time without retransmission requirements is at the core of RTP, whereas packet loss can be compensated by error-correction algorithms.

8.2 User Interfaces

Interfaces for computer instruments come in various forms, from the standard desktop graphical user interface (GUI) to the touchscreens of mobile devices and custom-built hardware controllers. In this section, we will examine some of the possibilities for building interactive controllers for music performance.

8.2.1 WIMP paradigm

The fundamental means of GUI structuring on standard OS platforms is based on the windows-icons-menus-pointer (WIMP) paradigm [95]. As the name implies, it depends on graphical containers called *windows*, which can be represented by *icons* and include lists of commands called *menus*, with all the GUI components manipulated by a *pointer*. Associated with this is the concept of a *widget*, which is the basic interface component, a window in generic terms, but having different types of graphical representation. Examples are push buttons, tick boxes, sliders, text windows, pull-down menus, etc. The design and functionality of such elements have matured significantly since they were first introduced and then popularised through the various desktop systems.

The WIMP paradigm is implemented by different computer OS platforms through their graphics subsystems and libraries. While the appearance and also some of the functionality might be system dependent, there is enough commonality between the various implementations to allow cross-platform development. This is provided in the form of frameworks and libraries in various programming languages. For example, Faust stand-alone programs can make use of the Qt or GTK libraries to build GUI-based applications. In this case, the same UI functions that we have used to provide control input when embedding programs in Csound or when using OSC are responsible for creating graphic controls. The Faust compiler will map controls to widgets according to the type requested (sliders, buttons, etc.). The example in Listing 8.5 will generate a GUI as shown in Fig. 8.1. Similarly, Csound has a number of GUI-based integrated development environments (IDEs) based on such libraries (e.g. CsoundQT, Cabbage, and Blue).

In the case of Python, a widely-avaliable cross-platform UI is Tkinter, which has already been encountered in Sect. 2.4. We will use it now to explore the basic concepts of UI design for computer instruments.

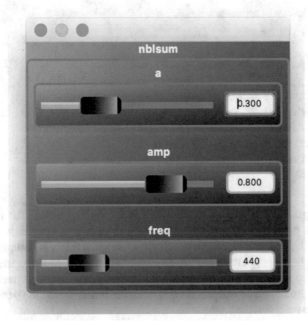

Fig. 8.1 A Qt-based GUI for a Faust stand-alone program based on Listing 8.5.

Behind the scenes in a GUI-based program there is a fundamental piece called the *event*, or *message*, loop. This is what keeps the software running, waiting for user input, which comes in the form of messages from the OS to the programs, triggered by user events. For instance, a program that displays a window receives messages related to it when the user clicks on, types in, moves, etc. that object. For this to happen, the OS needs to know to which program each window belongs, and whether tits is a sub-window of another one. This implies a certain hierarchical structure in the UI of a program, involving *parent* and *child* elements.

The OS will also have a standard set of message/event types that will be issued according to user actions. The program can select which ones it will respond to, providing the functionality for this. Associated with an action, we might have specific data. For instance, a mouse click will include the screen coordinates (x,y)[3] of the point where the action occurred; text typed into a given window will be passed as a string to the program; and so on. The information that is sent from the OS to the program will of course depend on the event type.

Turning to tkinter, let's build a user interface that will provide examples for the concepts developed so far. An application UI is defined by a user-provided class that is derived from `tkinter.Frame`. This means that, to build the UI, we need to specialise functionality that already exists in this base class. So a minimal program consists of:

---

[3] Horizontal coordinates are oriented from left to right and vertical ones, from top to bottom.

Listing 8.7 Minimal UI application.

```
import tkinter

class Application(tkinter.Frame):
    def __init__(self,parent=tkinter.Tk()):
        tkinter.Frame.__init__(self,master)
        self.master = parent
        self.master.title('Minimal Application')
        self.pack(padx=150,pady=20)
        self.master.mainloop()

Application()
```

This program might not do much, but it is a fully functional GUI application. It provides a main frame, derived from `tkinter.Frame`, a parent window (provided by `tkinter.Tk()`), from which the application frame is built, and an event loop (`mainloop()`). There are no contents, but in order to make the frame appear, we need to pad it with 150 pixels horizontally and 20 pixels vertically. The `pack()` is used to place widgets (in this case, the frame) inside the parent window and organise their positions within it.

There are various types of widgets that are provided by Tkinter. They share some basic attributes such as variables setting their graphical appearance, state, text, and functionality. A simple widget is `tkinter.Button`, which will allow us to demonstrate some of these common aspects:

Listing 8.8 Button UI application

```
import tkinter

class Application(tkinter.Frame):
    def click(self):
        self.n += 1
        self.button.config(text='clicked %d' % self.n)

    def createButton(self):
        self.n = 0
        self.button = tkinter.Button(self,
            text='clicked %d' % self.n,
            command=self.click)
        self.button.pack()


    def __init__(self,master=tkinter.Tk()):
        tkinter.Frame.__init__(self,master)
        self.master = master
        self.master.title('Minimal Application')
        self.pack(padx=150,pady=20)
```

```
     self.createButton()
     self.master.mainloop()

Application()
```

To structure the code more elegantly, we add a method to create the button, where we instantiate the widget and pack it. The widget will issue a callback when pressed, which we set to another class method that will change the button text. The definition of callbacks for window events is also known as *binding* an event to a function. In this case, we are doing the simplest type of binding, which is to respond to the basic widget action (clicking). However, there are other events that we could respond to. For instance, let's replace the button creation and callback with

```
def click(self,e):
  self.button.config(text='clicked (%d, %d)'%(e.x,e.y))

def createButton(self):
  self.button = tkinter.Button(self,text='click')
  self.button.bind(sequence="<ButtonPress>",
              func=self.click)
  self.button.pack()
```

In this case, the `click()` method is called an *event handler* for the pointer `<ButtonPress>` message. The callback is passed an extra argument, e, which is an `Event` object. In this case, we take the horizontal and vertical coordinates of the pointer and display them (Fig. 8.2). The `Event` class has a number of attributes that characterise the message, so we can use these in our application as sources of control data.

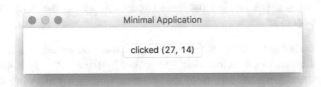

Fig. 8.2 A minimal application with a button and an event handler displaying the pointer coordinates.

The various other widgets provided by Tkinter follow a similar pattern of usage, with obvious variations due to their nature and application. However, a special type that deserves some special attention is `Canvas`. Objects of this type implement structured graphics and can be used as general-purpose widget builders. A canvas can hold a variety of user-drawn elements that can be handled independently. The *Shapes* application, discussed in Chap. 2 and presented in appendix B.1, makes use of this widget to create a 2-D space in which a geometric figure is moved about. We will use this example to outline how we can employ a canvas for interface building.

The following steps are used to build a UI based on `Canvas`:

1. Create and pack the canvas:

```
self.canvas = tkinter.Canvas(self,
                             height=self.size,
                             width=self.size,
                             bg="violet")
self.canvas.pack()
```

2. Create a figure in the canvas, in this case a circle (using `create_oval`):

```
circle = self.canvas.create_oval(self.size/2-10,
                                 self.size/2-10,
                                 self.size/2+10,
                                 self.size/2+10,
                                 fill="black")
```

3. Bind methods to events, using the `tag_bind()` method of `Canvas`. In this case, we respond to pointer button presses, pointer motion with button 1 depressed, and pointer button releases:

```
self.canvas.tag_bind(circle,"<ButtonPress>",
                     self.play)
self.canvas.tag_bind(circle,"<B1-Motion>",
                     self.move)
self.canvas.tag_bind(circle,"<ButtonRelease>",
                     self.stop)
```

4. Implement the event handler methods (`self.play()`, `self.move()` and `self.stop()`).

Finally, another noteworthy aspect of user interface frameworks such as Tkinter, is that they allow us to set callbacks to be invoked by the event loop at given times. These can be used to update aspects of the graphical interface (redrawing, displaying text, etc.), or for other applications (e.g. to trigger control messages to the synthesis engine at given time). In Tkinter, this functionality is implemented by the `Tk()` method `after()`, which fires a callback after a given timeout in milliseconds.

As a complete example, we demonstrate a simple MIDI controller with eight buttons covering a range of MIDI notes (Fig. 8.3). In this case, since we have several widgets of the same type, acting on different data (the MIDI note number), we need to find a way to pass the correct data to the event handler, according to the button pressed. The simplest way is to use a *lambda*, which is an anonymous function that will call the handler with the correct note number:

```
f = lambda e: self.note(e,n)
```

where `self.note()` is the handler method and n is the note number.

MIDI is played monophonically: each note on switches the last note off. We have to make sure that when the application is closed, no notes are left hanging. For this, we implement a handler for the WM_DELETE_WINDOW protocol of the parent window, which sends a MIDI note off message and destroys the window, closing the application. The rest of the code is self-explanatory and we use the mido package to provide the MIDI functionality. This instrument can be used to control the Csound and Faust MIDI synthesiser examples given in Sect. 8.1.1.

Listing 8.9 MIDI keys application

```python
import tkinter
import mido

class Application(tkinter.Frame):
    def note(self,event,num):
     if self.on > 0:
        self.o.send(mido.Message('note_off',
                                 note=self.on,
                                 velocity=90))
     self.o.send(mido.Message('note_on',
                              note=num,
                              velocity=90))
     self.on = num

    def createButton(self,n):
     button = tkinter.Button(self,text='%d' % n)
     f = lambda e: self.note(e,n)
     button.bind(sequence="<ButtonPress>",func=f)
     button.pack(side='left')

    def quit(self):
      self.o.send(mido.Message('note_off',
                               note=self.on,
                               velocity=90))
      self.master.destroy()


    def __init__(self,master=tkinter.Tk()):
     tkinter.Frame.__init__(self,master)
     self.master = master
     self.master.title('MIDI Keys')
     self.pack(padx=20,pady=20)
     self.notes = [60,62,64,65,67,69,71,72]
     for i in self.notes:
         self.createButton(i)
```

```
  self.on = 0
  if len(sys.argv) > 1:
   self.o = mido.open_output(sys.argv[1])
  else:
   self.o = mido.open_output()
  self.master.protocol('WM_DELETE_WINDOW',
                          self.quit)
  self.master.mainloop()

Application()
```

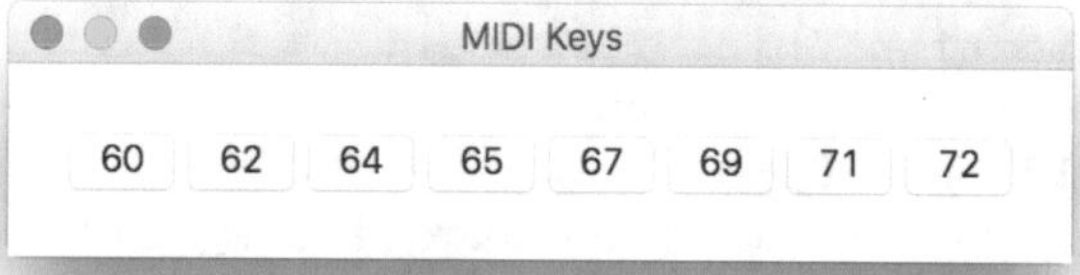

Fig. 8.3 A simple MIDI controller application.

In addition to the standard graphical components that can be manipulated via a pointer device, we also have to consider the use of the computer keyboard as another type of user interface input. An obvious use is to map keys to sound-producing functions, for example NOTE messages. More complex applications may try to discern patterns of interaction from the keyboard and pointer to provide further means of control.

8.2.2 Beyond WIMP

The successful aspect of WIMP has been in standardising a number of GUI elements, as well as simplifying them to work with basic means of input (e.g. pointer, buttons, and keyboard). A converse side to this is that the paradigm has been seen as increasingly limiting. For instance, it is tied to two-dimensional concepts and it does not take any notice of user feedback, so it is in fact, fairly limited in terms of interaction possibilities. Following its introduction and widespread adoption, it has attracted some criticism for these and other limitations [7, 28, 73].

A significant development in interactive interfaces is represented by the widespread use of hand-held computing devices such as mobile phones and tablets. In these, the WIMP paradigm has been expanded to take advantage of the many new gestures that can be captured from a user. The multi-touch screen, where the position and size of individual touches can be measured, processed and dispatched to an application, represents a step away from the use of a single pointer plus optionally the keyboard. In addition, often other, non-graphical types of control are possible, for instance by tracking the position and motion of the whole device. Among these we could mention the gyroscope, which provides a 3-D movement measurement, and the Global Positioning system (GPS), which can give the device's geographic location. All of these novel interface components make hand-held computing units a formidable platform for performance control. Some of these units also include a good capability for signal processing, which can turn them into a complete computer instrument.

8.2.3 Custom hardware

Riding the wave of the Maker movement [24], a number of computer musicians have started developing their own custom-built hardware interfaces, based on the increased availability of microcontrollers. The best example of these are the ones created with the Arduino development board [39]. The Arduino appeared as an open-source design, a lower-cost alternative to existing commercial boards such as the BASIC Stamp [81]. The original board consisted of an Atmel [4] 8-bit AVR microcontroller plus other components that facilitate programming and integration.

Alongside its range of development boards, the Arduino provides an IDE based on the AVR C/C++ compiler, together with Wiring library [74]. Applications, called *sketches* are written in the C language with the support from the various facilities provided by Wiring. Input and output (I/O) operations can be done with very small programs that are easily uploaded to the board from a host computer. The board layout allows direct access to the I/O terminals (general-purpose IO, GPIO) provided by the microcontroller, which allows them to be connected to sensors, switches, displays, and other hardware components.

The GPIO pins in the Arduino hardware provide three types of connections:

1. **Digital inputs**: these sense two levels of voltage (0 or +5 volts) producing 0 and 1 values (low, high) for the microcontroller.
2. **Digital outputs**: these produce two levels of voltage (0 or +5 volts) from a low or high signal sent by the microcontroller.
3. **Analogue inputs**: capable of sensing a number of voltage levels in the input (also between 0 and +5 volts), these are converted by an ADC at a given precision (e.g. 12 bit). Various types of components, such as potentiometers, light and movement sensors, etc., can be used with these to provide continuous parameter control interfaces.
4. **PWM waveform**: analogue output is provided as a pulse-wave modulation waveform, suitable for driving small electric motors and similar components.

Alongside the Arduino, a number of pluggable boards, known as *shields* can be used to extend the functionality, providing for instance proper DAC facilities, MIDI integration, data card access, wireless networking, etc. Generally speaking, Arduino-based hardware is not suitable for audio signal-processing tasks, but can however, be used to construct various types of physical interfaces to computer instruments running on a host platform[4]. Communications can be made through serial connections (e.g. USB) or standard IP networks, using ad-hoc messaging, or established protocols such as MIDI and OSC. Following the success of the Arduino, various clones and similarly-designed development boards have been made available, supporting the emergence of a new field of performing technology called *physical computing*.

8.3 Conclusions

Interaction is a very wide topic, and is fundamental to the design of computer instruments. In this chapter, an overview of the area was provided, attempting to touch on all relevant aspects of the subject. We saw how the methods of communication between devices and software play a central role, providing the binding between different parts of a complex system. Traditionally, MIDI would have been used as a de-facto standard for controlling musical instruments, but with the emergence of OSC, it has been superseded, at the least as far as the technology is concerned. Likewise, WIMP has been the major paradigm for the building of software user interfaces, but it is being supplanted by various new ways of interacting with computing devices. The possibilities that custom hardware offer to computer instrument interaction design are also a very important development that has been added to the mix. It is therefore safe to say that the field of user interfaces is in constant development and we should expect increasingly powerful capabilities to be available to computer music practitioners in the immediate future.

---

[4] It is the case, however, that some recent models, such as the Arduino Due, have a processor that has some capabilities for synthesis and processing

Chapter 9
Computer Music Platforms

Abstract In this chapter, we will explore the various platforms for computer music that are available to practitioners today. The discussion will be steered mostly towards general-purpose computing, as this has proved the more enduring form of such platforms. We will look closely at the software and hardware environments available for desktop, mobile and web applications, discussing also the emergence of the DIY platform, and the concept of an Internet of Musical Things.

Computing platforms come in all shapes and sizes, with varying capabilities and applications. Some of them are general purpose: they can be applied to a wide range of problems but will not be particularly adapted to specific classes of them. These platforms are represented typically by desktop and laptop machines[1], but in some ways we can say that other platforms such as mobile devices (phones and tablets) also have increasingly general-purpose characteristics. Some small computing devices that can be embedded in a larger system can also be considered in this category, especially if they run under desktop operating systems such as Linux.

General-purpose platforms provide a stable development and performance environment for computer music. It is interesting that they actually allow programs that were developed almost fifty years ago to continue to be used if needed. For example, Max Mathews' MUSIC V [65] program can be employed in modern systems because of the availability of FORTRAN compilers that can take the original code (with little modification) and produce working software. This longevity allows the ideas and concepts that have emerged during the long history of computer music to be retained and applied, providing a continuous line of development.

Dedicated computing platforms are designed to be specific to certain problems. These include what are often called *digital signal processors* (DSP computers), which have built-in hardware facilities for certain types of operations. The main justification for them is that they can speed up significantly some critical parts of the computation of a problem. However, this is not always the case, as general-purpose computers become increasingly more powerful. Additionally, they generally require

---

[1] It has become usual to lump all of these under the *desktop* description, even though some are portable and others not.

© Springer International Publishing AG 2017
V. Lazzarini, *Computer Music Instruments*,
https://doi.org/10.1007/978-3-319-63504-0_9

code to be tailor-made for specific hardware capabilities, and this may become obsolete as new versions of the platform are introduced. Programs written for general-purpose computers are much more resilient against obsolescence, particularly if they are written in a widely-used, well-maintained, cross-platform language.

Two types of dedicated hardware, however, appear to have overcome these difficulties to become established computing platforms for the longer term: physical computing devices (such as Arduino boards), discussed in Sect. 8.2.3; and graphic programming units (GPUs). It is true that both of these have a wide range of applications and in some sense share general-purpose qualities. Arduino boards, for instance, can be applied to several problems (including, depending on the processor, simple audio processing). The more successful GPU hardware has actually been re-named *general-purpose* GPUs (GPGPUs), given the fact that they are being used beyond their original application area (graphic displays). Another important point is that both of these dedicated platforms have been given stable cross-platform programming environments (with high-level languages). Although these environments are specific to them, they allow programmers to maintain a good deal of compatibility between different versions fairly easily.

In the following sections of this chapter, we will examine the more salient characteristics of a variety of sound and music computing platforms. Most of these platforms tend to be fairly general purpose, but some specific capabilities will also emerge from the discussion, allowing users to consider their choices when implementing computer instruments.

9.1 Desktop

The desktop platform appeared with the introduction of microcomputer central processing units (CPUs), It was preceded by larger-size machines such as mainframes and minicomputers. In terms of function as a general-purpose platform, we can perhaps trace a direct line from the IBM 704, used by Mathews to write and run his MUSIC I program [45], to today's desktop, passing through machines such as the PDP-11 minicomputer. There are important differences, however, between each of these, especially in the practicalities of their operation. A mainframe in the 1960s operated in batch mode, that is, programs were scheduled to run in a sequence amongst others from various users. The computer was fed with a series of punched cards or paper tape. These were prepared using special teleprinter machines[2]. The outputs would be written to a paper printout and magnetic data tape. In the specific case of computer music, for the synthesised sound to be played back, the magnetic tape would be transferred to a dedicated machine containing a digital-to-analogue converter, and then recorded in an analogue form (usually a magnetic audio tape).

---

[2] Control of operation would also use these (as well as console switches). Some computers would allow online connections using teleprinter terminals, but this was not generally accessible to the ordinary user.

Minicomputers allowed users to get closer to machine operation. Since these were smaller systems, teletype terminals were more commonly present, first with printed paper output, and later on with video display units (VDUs). Operating systems started to become more interactive and, with the appearance of the UNIX OS, a multi-user time-sharing environment started to become the norm. Development tools also started to crystallise around the C language, some utility programs (such as `make` and text editors) and the UNIX system. Some computer music laboratories started to develop site-specific software environments that could eventually be ported to other similar platforms, such as the CARL installation [62].

By the time microcomputer CPUs became more generally available, scientific computing and software development were increasingly based on graphical workstations with WIMP interfaces (based on the X Window System standard and UNIX). This provided, for those who could afford it, a very stable and feature-full computer music environment. At the same time, the first consumer-oriented microcomputers were already enjoying some popularity with hobbyists and early business users [1]. These machines would include the original Apple II machine, the first IBM personal computers, and a host of 8-bit BASIC-language oriented micros. The arrival of the Macintosh with its GUI spawned considerable interest, with a number of early computer music software packages appearing, mostly dealing with MIDI data. Interestingly, due to a lawsuit with the recording company of the same name, Apple was not allowed to include built-in MIDI interfaces in its computers, opening up the market for some competitors, such as Atari, who did include them.

By the early 1990s, consumer desktop computers were able to do off-line sound synthesis using some of the C-language packages ported from UNIX. MS DOS-based PCs and Apple Macintoshes had Csound, and the Atari had both Csound and Cmusic, as well as a number of interesting processing programs developed by the Composer's Desktop Project (CDP). For institutions, the UNIX workstation was the chosen platform, but these were not generally within the reach of the ordinary user. Audio hardware in the form of DACs/ADCs started to become available, first as dedicated outboard devices, and then more commonly as part of the multimedia desktop. Somewhere around the end of the decade, PCs and Macintoshes became more widely capable of real-time audio synthesis and processing, with WIMP GUIs as the norm.

The modern desktop platform is a complete computer music development, production and performance environment. It allows users to navigate through a range of software environments, combine them by chaining operations, by mixing languages, displaying data, setting up controls, and generating audio of very high quality in real time (as well as offline). This book, as well as the concepts, the software, the graphic plots it contains, and even its typesetting, is a testimony to this. The development of the desktop platform as an environment for computer music has depended on some key components, which we outline below [45]:

1. **The MUSIC N paradigm**: this is is an emergent set of properties arising from music programming systems such as Csound (and to some extent, Faust). Its essence is that a flexible platform for computer music should be programmable. The principle of a compiler for instruments, which harks back to MUSIC III [63],

and its role of translating a symbolic representation of the synthesis signal flow as a DSP operation graph, is the most significant aspect of this paradigm

2. **Object-oriented programming**: the MUSIC N paradigm, from a programming point-of-view, is realised through the principles of object-oriented programming. The idea of instrument classes, instantiated in performance, which are composed of instances of unit generators (opcodes and DSP programs) is a very powerful and general principle. We can therefore talk of signal-processing objects at various levels, with various descriptions in terms of attributes and a common operation (audio synthesis or control, depending on the type).

3. **The multi-language paradigm**: as we have been constantly reminded by the various examples in this book, at the centre of computer instrument development we find the multi-language design paradigm [78]. The intermixing of a DSL, a scripting language, and sometimes a system-implementation language (e.g. C/C++) provides a very flexible and complete set of solutions for sound synthesis and processing (and music making, sound design, research, etc.). It allows large design problems to be broken down into separate components, through the principle of separation of concerns [19].

4. **Free and open-source software**: one of the main mechanisms that made some of the main developments in computer music under the desktop platform has been the availability and exchange of source code. The most seminal pieces of software in the area, including all MUSIC N systems and descendants, have been available as open-source software, even before the concept was fully developed. This cross-fertilisation of ideas not only allowed the emergence of the desktop and other computing platforms as viable platforms for music making, but also helped the establishment of computer music as a respected research and artistic discipline. A remarkable example of this success is the Linux OS and the wide range of audio and music software provided for it [82].

5. **Ecosystems for music making**: software ecosystems are a group of packages that function in a cooperative and complementary way on the same or related platforms. In computer music, these take the form of digital audio workstations (DAWs), plugin systems (such as VST and LADSPA), audio and MIDI plumbing infrastructure (i.e. Jack connection kit and system libraries), and music and general-purpose programming environments. Such ecosystems have allowed users to approach the desktop platform from different perspectives, from various levels of technical expertise, through the many *entry points* that they provide (Fig. 9.1).

The concept of ecosystems for computer music is analogous to the principles of the multi-language paradigm, based on the complementarity of different pieces of software. In the modern desktop platform, it is very common to start the work with one type of application, and then begin to integrate others as the demand for other specific functionalities arise. In Fig. 9.1, we also see how these applications can be connected in an ecosystem. For example, a music DSL can be used for plugin development, producing plugins of a certain specification that are loaded by a DAW. A computer instrument, designed with a music DSL can become a software synth that is controlled via MIDI or OSC from an external controller or, again, from a DAW.

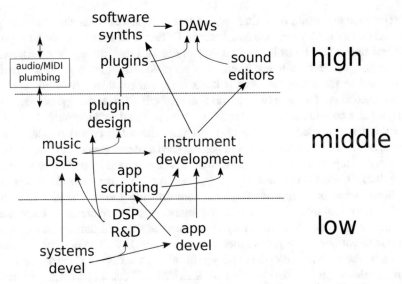

Fig. 9.1 Desktop computer music ecosystem and the three levels of interaction with software tools (low, medium, and high).

Similarly, we might use the instrument to produce an audio file that we manipulate with a sound editor and then load into a DAW project. Progressing downwards to a lower level, demanding further technical skills, we might develop algorithms that would be integrated into a music DSL for general use. Gluing the ecosystem together, we also have audio and MIDI plumbing software that allow applications to talk to each other and to external devices (MIDI controllers, network, DAC/ADC, etc.). The desktop platform provides an accessible and affordable environment for music and sound making that is quite formidable in its wide range of application possibilities.

9.2 Mobile

In the past decade, mobile computing platforms have emerged as an alternative to desktops. Early experiments in porting some music systems to these were limited significantly by the lack of processing power and of flexible tools for development. For example, it used to be the case that these devices did not have hardware floating-point units (FPUs) and so signal-processing operations needed to be programmed using fixed-point arithmetic in order to run efficiently. Porting large systems such as Csound was impractical. However, as this platform developed, it became possible to provide software that went beyond simple proof-of-concept demos. Nowadays, mobile devices are very capable of being used in performance, composition and sound design.

There are two major operating systems that fully support computer music applications on mobile devices, iOS and Android. Of these, the former has better support for real-time audio IO as it provides system libraries that, through good hardware integration, can deliver lower levels of latency. Android took a while to provide good support for audio, but recently it has improved considerably. Music application development for these two platforms depends heavily on the system-supported languages, a combination of Java and C/C++ for Android, and Objective C for iOS. For this we also need a desktop host machine with a full cross-compilation tool chain, as mobile platforms are not in general used as development environments.

The desktop computer music ecosystem can be extended by mobile applications (Fig. 9.2). Often it is the case that an application might have a mobile version that replicates some of its functionality or is complementary to it. The idea of mobile DAWs is an example of this. While they cannot possibly re-create a complex environment on a mobile device, they allow some editing and music making *on the go* that might also share project files with a desktop DAW. The presence of music DSLs on the mobile platform is also significant. In the case of Csound and Faust, this translates into the development of dedicated self-contained *digital music instruments* (DMIs)[3] that are based on them as audio engines, and use post-WIMP mobile interfaces for interactive control. To facilitate this, application development makes use of application programming interfaces (APIs) that are available to mobile platforms, such as the Csound (CsoundObj) API and other system-supplied ones. Csound is also present in the form of complete mobile-IDE applications that partly replicate the computer instrument development environment found on desktops.

9.3 Web Browsers

More recently, web browsers have become a kind of virtual platforms in their own right, working under the various OSs and hardware supported by them. In some cases they have become the OS (as examples, we have ChromeOS and FirefoxOS, derived from their homonymous browsers). The web platform, as we can more generally call it, is delivered by different vendors, but is unified by common standards. These, at the time of writing, are HTML 5, and as the system language, Javascript (JS). There exist further browser-specific facilities, but these are not guaranteed to work universally. Applications are delivered through the hypertext transfer protocol (HTTP, and its secure version HTTPS) as plain text that is interpreted by the browser engine. They can avail of network connections as well as local facilities, such as disk and device access, mediated through a secure interface.

Web browsers extend the computer music ecosystem in important ways (Fig. 9.2). As they are agnostic with regard to the underlying software and hardware platform and can be adapted to a variety of them, they provide an environment where

---

[3] While we could argue that a platform such as the desktop is a DMI, or that the computer instruments we have been discussing in this book are DMIs, we should reserve the term for specific self-contained applications that make use of specific hardware interfaces for performance.

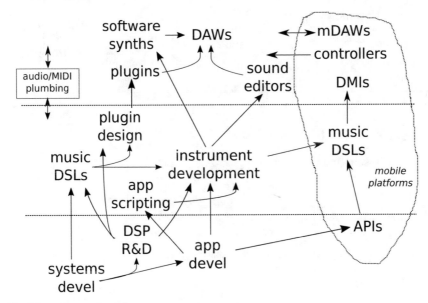

Fig. 9.2 Desktop and mobile computer music ecosystems.

the local and remote spaces meet. Applications are provided from outside locations but are run in the users' machines. There is a wide range of software types operating at different levels: web DAWs (wDAWs), software synths (also DMIs), plugins, remote controllers, etc. Again, the presence of music DSLs is significant. The most common way that this is facilitated is through JS cross-compilation of code bases that are either in C or C++. In the case of Faust, its compiler can generate JS application code, plus an accompanying HTML user interface. For Csound, there are actually two in which ways the system is present on the platform: cross-compilation of the engine into JS and as a portable Native Clients (PNaCl) frontend (with a JS interface) [46]. The latter is very efficient and provides excellent near-native performance, but it relies on support for PNaCl in the browser (currently only Chrome and Chromium support is widely available).

Web JS applications are interpreted by a browser engine, which translates the code into the actual computational actions to be performed. While the JS engines produce very efficient results, these are not as good as in natively-implemented programs. So there is a penalty to be paid for applications to be present on the web platform. Additionally, JS audio code relies on the Web Audio API, which is a collection of JS objects supporting audio processing on the platform, called *nodes*. Some of these nodes replicate parts of music DSLs, as they implement oscillators, envelopes, filters, etc natively and so they are very computationally efficient. Web Audio also provides the audio plumbing to the host OS and this is the element that is heavily used by DSLs such as Faust and Csound. Music DSLs are placed in a special Web Audio object called a *script processor* node, which is used to run pure

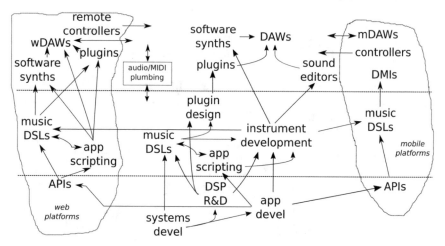

Fig. 9.3 Web, desktop and mobile computer music ecosystems.

JS code. While it provides the required functionality, this mode of operation has a number of issues that have not yet been addressed, namely lower priority execution (vis-a-vis the native nodes), higher IO latency, and generally an odd integration with the rest of the Web Audio API [60].

PNaCl applications are compiled directly from C/C++ code, and thus Csound has been ported directly to this environment. In order to communicate with the browser, the code uses a JS interface, which in the case of Csound, implements a front-end control allowing users to operate Csound inside a browser and construct HTML interfaces for it. This allows the development of highly efficient applications, as well as fully-capable web-platform IDEs for computer instrument design. Since all the heavy computing happens at near-native speed, there is very little penalty for switching to a web platform in this case. PNaCl applications do not depend on Web Audio and use the C++ Pepper API for sound IO.

The web platform makes the relationship between the local and the remote computing spaces very fluid. It allows the supply of audio applications over a wide area with no real obstacles. Since it is deployable over a heterogeneous set of hardware bases, it opens up the possibility of distributed audio computing with local control and output. It has been used for concerts and installations where the audiences' mobile devices are used as components in a large-area interconnected macro computer instrument. It allows networks to be used as links between mutually-dependent units in ways that we have not encountered before.

9.4 DIY Platforms

Intersecting all of these, desktop, mobile, and web, we find the DIY platform. Inspired by the Maker movement [24], this shares elements with all three in that it is generally based on what we would class as a desktop OS environment, it can be ultra mobile, and can avail of the web and the internet application ecosystem. This platform is quite heterogeneous in terms of hardware, but is generally homogeneous in terms of software base, firmly grounded on Linux free and open source tools. It is characterised by a multi-component, relatively inexpensive set of computing hardware, which might include a full computer based on a low-power CPUs (such as the ARM-based Raspberry PI and Beaglebone, or the i386-based Galileo and Edison; see Fig. 9.4), solid-state storage, HDMI (or sometimes plain VGA) displays, external or internal audio DAC/ADC, MIDI, ethernet and wifi networks, and a myriad of types of electronic controls, sensors, and displays, sometimes attached to auxiliary processing in the form of Arduino or other interface boards (Fig. 9.5).

The DIY platform has evolved from a long history of hobby electronics that stretches all the way back to the end of the Second World War. The general-purpose computing side of it was spurred on by projects such as One Laptop per Child, which aimed to provide an affordable platform for learners. The connection with education is very strong, as some of the hardware has also targeted the teaching of programming languages. In general, the DIY platform is very akin to an experimental laboratory, as it embodies the qualities of reconfiguration, development of specific solutions, and combination of re-purposed components.

While there is no specific shape to the DIY platform, some elements are ubiquitous: the aforementioned Linux operating system, scripting environments (such as Python and the Wiring library), and music DSLs. The latter play, one more time, a very significant part in the platform, supporting the development of computer instruments. They allow not only similar approaches to desktop music making, but also the development of custom fully self-contained DMIs, which can be created for specific or more general applications. In that scenario, a music DSL such as Csound becomes a general-purpose audio engine for DMI design.

The impact of Free and open-source software is significant here. These support the dissemination of ideas and the sharing of code, giving the platform an infrastructure that allows new processes of creation to flourish. This is evident in computer music software for custom devices, where the most successful packages stem from established Free and open-source communities. Many of these offer features that go beyond what is available in off-the-shelf systems. The proliferation of makers, allied with high-quality development tools and a creative user base, has helped develop the custom hardware platform as a featureful environment for computer music.

Fig. 9.4 Two CPU boards commonly used in DIY platforms.

9.4.1 The Internet of Musical Things

A significant aspect of the DIY platform is how devices can be created to take advantage of the internet infrastructure to be connected over a wide area. Application examples are represented by small computing devices that can be remotely controlled and used to gather input data, to generate audio output and to interact with people and the environment. We can think of extended computer instruments acting across a large space, which might possibly even be of geographical significance. As discussed in Sect. 9.3, this approach facilitates the distribution of computing load between a variety of units. More importantly, it does not necessarily require a dedicated networking infrastructure, instead allowing us to ride on the ubiquitous

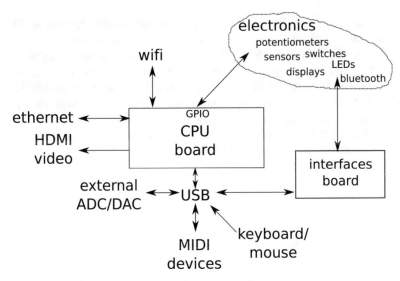

Fig. 9.5 The DIY platform as a collection of interconnected components.

presence of the internet. The emergence of this Internet of Musical Things will possibly continue to affect the way artists and audiences interact with music-making devices. It opens an even wider range of possibilities for multimedia integration of various creative concerns and for continued support of ubiquitous music activities [36], particularly with regard to the design of computer instruments.

9.5 Conclusions

As we have seen, there is a wide range of platforms available for the design, development, implementation, and performance of computer music instruments. Some of these have only appeared in the last few years and have yet to reach a stable form, while the desktop is perhaps a mature environment, whose foundations can be traced all the way back to the beginnings of computer music. In the last twenty to twenty-five years, this platform has evolved into a formidable space for sound and music computing. It is very hard to predict how things will develop in the years to come as the consumer market for computing devices is changing continuously and the same conditions that led to the availability of the desktop as a viable computer music platform might lead to the development of other, more advanced platforms. It is, however, hard to see how we can replace the ability to cover the research, experimentation, implementation and performance tasks that are currently managed by the desktop. Mobile devices are not yet fully capable with respect to development of applications on their own; web browsers will still need some vestige of a general-purpose OS underneath them to manage hardware and IO.

Predicting the future is a perilous task, but we can be sure that things will not stay the same. Of all the chapters in this book, this one is possibly the only one that is in danger of become dated in a very short space of time, as computing platforms evolve. It might, however, serve as a good testimony to a given point in time. Today, almost sixty years on from its birth, we find computer music as an established discipline, for which a mature platform, as well as various other exciting alternatives, is available to support the work of composers, sound designers, researchers, and educators. We can look forward to a future of continued development from the very solid foundations that have already been laid.

Appendices

Appendix A
Signal Processing Mathematics

This appendix provides some support for the decoding of the mathematical expressions given in this book. We will look at the specific operations and concepts that are fundamental to the principles of synthesis and digital signal processing that we have developed and implemented in Part II.

A.1 Fundamental Concepts and Operations

In addition to the rules for the usual arithmetic operations $(+,-,\times,/)$, the following apply for *exponentiation*:

$$
\begin{aligned}
a^0 &= 1 \quad (a \neq 0) \\
a^1 &= a \\
a^x a^y &= a^{x+y} \\
\frac{a^x}{a^y} &= a^{x-y} \\
(a^x)^y &= a^{xy} \\
a^x b^x &= (ab)^x \\
a^{-x} &= \frac{1}{a^x}
\end{aligned}
\tag{A.1}
$$

From these definitions, we can work out the exponentiation of a number. For instance

$$
a^5 = a^{1+1+1+1+1} = a \times a \times a \times a \times a
\tag{A.2}
$$

The logarithm of a number is defined by

$$
\log_b a = x \quad \Rightarrow \quad a = b^x \quad (b \neq 0 \text{ and } b \neq 1)
\tag{A.3}
$$

where b is called the *base* of the logarithm, which is most commonly set to one of 2, 10, or $e = 2.718281828459...$ (the *natural* base of logarithms). In this book, log (with no subscript) implies base e.

From this and the definitions in Eq. A.1, we have the following relations:

$$\log_c ab = \log_c a + \log_c b$$
$$\log_c \frac{a}{b} = \log_c a - \log_c b \qquad\qquad\qquad \text{(A.4)}$$
$$\log_c a^b = b \log_c a$$

From these we can see that the ratio of two values becomes a difference of their logarithms. While a linear distance between two quantities a and b is defined as the difference $b - a$, their logarithmic distance is $\log b - \log a$. The former applies to our understanding of physical space, whereas the latter relates closely to the way we perceive amplitudes and frequencies.

A.1.1 Vector and matrix operations

The transposition of a vector is the exchange of columns for rows (and vice-versa):

$$[a\,b\,c\,d]^T = \begin{bmatrix} a \\ b \\ c \\ d \end{bmatrix} \qquad\qquad\qquad \text{(A.5)}$$

The product of two square matrices A and B is defined as

$$AB = \begin{bmatrix} a & b \\ c & d \end{bmatrix} \times \begin{bmatrix} e & f \\ g & h \end{bmatrix} = \begin{bmatrix} ae+bg & af+bh \\ ce+dg & cf+dh \end{bmatrix} \qquad\qquad \text{(A.6)}$$

Note that the commutative property does not apply here, so, in general, $AB \neq BA$ (although it is obviously possible to engineer cases where this is not true).

Likewise, we can effect the product of a matrix A of dimensions $N \times N$ by a column vector B of size N, producing a column vector also of size N:

$$AB = \begin{bmatrix} a & b \\ c & d \end{bmatrix} \times \begin{bmatrix} e \\ f \end{bmatrix} = \begin{bmatrix} ae+bf \\ ce+df \end{bmatrix} \qquad\qquad\qquad \text{(A.7)}$$

A.1.2 Sum and product

Shorthand forms for sum and product are defined as follows

$$\sum_{n=s}^{N} a_n = a_s + a_{s+1} + a_{s+2} + \dots + a_{N-1} + a_N \tag{A.8}$$

and

$$\prod_{n=s}^{N} a_n = a_s \times a_{s+1} \times a_{s+2} \times \dots \times a_{N-1} \times a_N \tag{A.9}$$

In these expressions, n is called the *index* (of the summation or product), and its range is normally explicitly defined (unless it can be deduced from elsewhere). Summation formulae can have *closed* forms, as discussed extensively in Chap. 4, the simplest of which is just

$$\sum_{n=1}^{N} n = \frac{N(N+1)}{2} \tag{A.10}$$

called an *arithmetic series*.

Some fundamental relations should also be noted:

$$
\begin{aligned}
\sum_{n=s}^{N} f(n) \pm \sum_{n=s}^{N} g(n) &= \sum_{n=s}^{N} [f(n) \pm g(n)] \\
\sum_{n=s}^{N} k f(n) &= k \sum_{n=s}^{N} f(n) \\
\sum_{n=s+m}^{N+m} f(n) &= \sum_{n=s}^{N} f(n+m) \\
\sum_{n=0}^{2N+1} f(n) &= \sum_{n=0}^{N} [f(2n) + f(2n+1)]
\end{aligned}
\tag{A.11}
$$

A.1.3 Polynomials and functions

Polynomials are expressions involving a variable raised to various powers:

$$\sum_{n=0}^{N} a_n x^n = a_0 + a_1 x + a_2 x^2 + a_3 x^3 + \dots + a_N x^N \tag{A.12}$$

The multipliers a_n associated with the various terms are called *coefficients*, and the *order* of a polynomial is given by N, its highest exponent. A function is an expression that maps one or more parameters into a value. A *polynomial* function

$f(x)$ is one that involves an expression such as the one in eq A.12, where an input x is mapped to the result of the polynomial defined by $f(x)$. Such functions can be evaluated for various ranges of x.

A.2 Trigonometry

An angle can be the defined as the space between two intersecting lines. It can be measured in degrees, ranging from 0 to $360°$, with $90°$ marking the angle between two perpendicular lines, called a *right* angle. A triangle with one right angle is called a *right* triangle. Angles can also be measured in relation to the diameter d and circumference c of a circle whose centre is the intersecting point of the two lines. This is the preferred way, and can be formed by this relation:

$$c = \pi d \tag{A.13}$$

So, for a circle or radius 1, called the *unit circle*, angles can be measured linearly on the circumference from 0 to 2π, with $\frac{\pi}{2}$ being equivalent to a right angle. This can be extended to all angle measurements by defining a scale on this basis, whose unit is called a *radian* (Fig. A.1). All angles in the signal-processing formulae in this book are thus defined.

The fundamental trigonometric functions are first defined in relation to a right triangle with sides A, O, and H, where the right angle is placed between the A and O sides (Fig. A.2). Firstly, from Pythagoras, we have

$$H^2 = A^2 + O^2 \tag{A.14}$$

In this arrangement, we have, for an angle θ set between the A and H sides,

$$\sin(\theta) = \frac{O}{H}$$
$$\cos(\theta) = \frac{A}{H} \tag{A.15}$$

To simplify the relationships, we set $H = 1$, making $\sin(\theta) = O$ and $\cos(\theta) = A$. In this case, we can see how the functions $\sin(x)$ and $\cos(x)$ taken together can track the position of a point on the unit circle. For instance, in Fig. A.2 the (x,y) coordinates of point b are $(H\cos(\theta), H\sin(\theta))$, with respect to the centre of the circle. Thus we have $\sin(0) = 0$, $\sin(\pi/2) = 1$, $\sin(\pi) = 0$, and $\sin(3\pi/2) = -1$; similarly, $\cos(0) = 1$, $\cos(\pi/2) = 0$, $\cos(\pi) = -1$, and $\cos(3\pi/2) = 0$. Note that

$$\cos(\theta) = \sin(\theta + \pi/2)$$
$$\sin(\theta) = \cos(\theta - \pi/2) \tag{A.16}$$

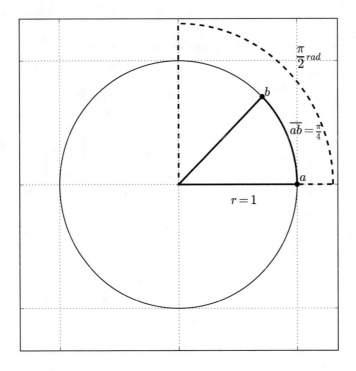

Fig. A.1 Angle definitions: for a circle of radius 1, the total circumference measures 2π, and the segment $\overline{ab}$ in the picture measures $\pi/4$. This is equivalent to an angle of $\pi/4$ radians. A right angle (dashes) measures $2\pi rad$.

and

$$\sin(\theta) = -\sin(-\theta)$$
$$\cos(\theta) = \cos(-\theta)$$

(A.17)

In addition to these functions, we can also define the tangent as

$$\tan(\theta) = \frac{O}{A} = \frac{\sin(\theta)}{\cos(\theta)}$$

(A.18)

which is equivalent to the length of the line $\overline{de}$ in Fig. A.2. To recover the angle from a given ratio of triangle sides, we have inverse functions:

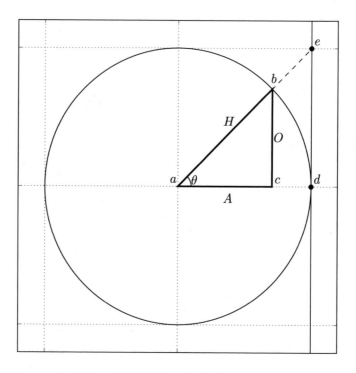

Fig. A.2 Trigonometric circle: a circle of radius H inside which we construct a right triangle abc, with sides A, O, and H. An angle θ is defined between H and A. A line that is tangent to the circle at point d is drawn, with which we can define another right triangle aed.

$$\arcsin\left(\frac{O}{H}\right) = \cos^{-1}\left(\frac{O}{H}\right) = \theta$$

$$\arccos\left(\frac{A}{H}\right) = \sin^{-1}\left(\frac{A}{H}\right) = \theta \qquad \text{(A.19)}$$

$$\arctan\left(\frac{O}{A}\right) = \tan^{-1}\left(\frac{O}{A}\right) = \theta$$

and $\cos(\arccos(\theta)) = \theta$ etc.

A.2.1 Identities

There are a number of identities between trigonometric functions that can be used to manipulate expressions using them:

$$\cos(\theta)^2 + \sin(\theta)^2 = 1$$

$$\cos(\theta)^2 = \frac{1 + \cos(2\theta)}{2}$$

$$\sin(\theta)^2 = \frac{1 - \cos(2\theta)}{2} \tag{A.20}$$

$$\cos(2\theta) = \cos(\theta)^2 - \sin(\theta)^2$$
$$= 1 - 2\sin(\theta)^2 = 2\cos(\theta)^2 - 1$$

$$\sin(2\theta) = 2\cos(\theta)\sin(\theta)$$

$$\cos(\theta \pm \phi) = \cos(\theta)\cos(\phi) \mp \sin(\theta)\sin(\phi)$$

$$\sin(\theta \pm \phi) = \sin(\theta)\cos(\phi) \pm \cos(\theta)\sin(\phi)$$

$$\cos(\theta)\cos(\phi) = \frac{\cos(\theta - \phi) + \cos(\theta + \phi)}{2}$$

$$\sin(\theta)\sin(\phi) = \frac{\cos(\theta - \phi) - \cos(\theta + \phi)}{2} \tag{A.21}$$

$$\cos(\theta)\sin(\phi) = \frac{\sin(\theta + \phi) - \sin(\theta - \phi)}{2}$$

$$\sin(\theta)\cos(\phi) = \frac{\sin(\theta + \phi) + \sin(\theta - \phi)}{2}$$

$$\cos(\theta) + \cos(\phi) = 2\cos\left(\frac{\theta + \phi}{2}\right)\cos\left(\frac{\theta - \phi}{2}\right)$$

$$\cos(\theta) - \cos(\phi) = -2\sin\left(\frac{\theta + \phi}{2}\right)\sin\left(\frac{\theta - \phi}{2}\right) \tag{A.22}$$

$$\sin(\theta) \pm \sin(\phi) = -2\sin\left(\frac{\theta \pm \phi}{2}\right)\cos\left(\frac{\theta \mp \phi}{2}\right)$$

A.3 Numeric Systems

In this book, we use three numeric systems that have some important characteristics. The first one of these is the *integers*, which represent whole-number quantities, both positive and negative. These are used mainly when counting, keeping track of summation indices, delay time in samples, etc.

The next system that we employ is the *real* numbers. These represent the one-dimensional continuum, for instance time passing from one instant to another, infinitesimal differences in values, etc. We can think of these numbers as points on a line that extends from $-\infty$ to ∞, where a number a is to the left of b if $a < b$, and to the right of it if $a > b$. Between a and b lie an infinite number of points. The reals also include some special numbers such as π and e, as well as quantities such as $\sqrt{2}$.

This system encompasses all integers and all fractions of integers (which are also called *rationals*). The synthesis and processing theory developed in Part II depends heavily on it.

The third system that we use extends the reals to two dimensions. These are called *complex* numbers. We will outline its key characteristics here. A complex number is defined as a conjunction of two real numbers (we can think of it as a *tuple*) a and b, where one of them carries a special mark, j, to distinguish it from the other:

$$z = a + jb \tag{A.23}$$

This special symbol has an algebraic interpretation, $j^2 = -1$, which we will use when necessary. For the moment, we can think of it as something that marks b as different from a, placing it in a different dimension. So if a lies on one line, b lies on a different one that is perpendicular to it (Fig.A.3)

This is the geometric interpretation of j, placing the quantity b in that direction. The number z then lies at the intersection of a line starting from a at a right angle to the real line and a line from b parallel to the real line. So it is a two-dimensional quantity, defined on a plane, and requiring a and b to locate it. The line on which b lies is called the *imaginary* line and, taken together, this and the real line form a set of axes that divides the plane (the *complex plane*, shown in Fig. A.3) into four quadrants $(+, +j; -, +j; -, -j;$ and $+, -j)$.

Complex arithmetic follows the usual rules, so we have the following relationships for two complex numbers $z = a + jb$ and $w = c + jd$:

$$z + w = (a + jb) + (c + jd) = a + c + j(b + d) \tag{A.24}$$

$$z - w = (a + jb) - (c + jd) = a - c + j(b - d) \tag{A.25}$$

$$z \times w = (a + jb)(c + jd) = ac + j^2 bd + j(bc + ad) \tag{A.26}$$

Since we have defined earlier that $j^2 = -1$, then

$$z \times w = ac - bd + j(bc + ad) \tag{A.27}$$

$$\frac{z}{w} = \frac{a + jb}{c + jd} = \frac{(a + jb)(c - jd)}{(c + jd)(c - jd)} = \frac{ac + bd + j(bc - ad)}{c^2 + d^2} \tag{A.28}$$

An alternative representation for a complex number is through its *polar* form. To get this, we use a trigonometric interpretation of the complex number, where

$$\Re\{z\} = A\cos(\phi) \tag{A.29}$$

and

$$\Im\{z\} = A\sin(\phi) \tag{A.30}$$

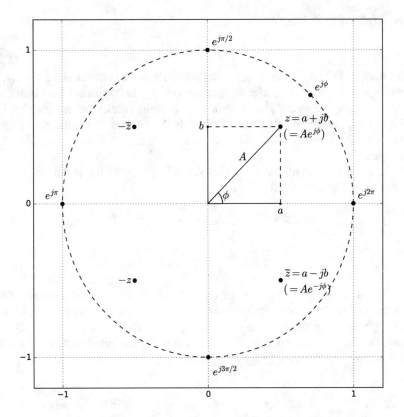

Fig. A.3 The complex plane: a complex number $z = a + jb$ may lie anywhere in this plane. In this particular case, z is 0.5 + 0.5j, and its coordinates are $(a,b) = (\Re\{z\}, \Im\{z\}) = (0.5, 0.5)$. Its complex exponential form is $Ae^{j\phi} = \sqrt{2}e^{j\pi/4}$. Its complex conjugate $\bar{z}$ is also shown, along with $-z$ and $-\bar{z}$. The figure also demonstrates that a number e^{jx} will always lie on the unit circle (dashes). The numbers $e^{j\pi/4}$, $e^{j\pi/2}$, $e^{j\pi}$, $e^{j3\pi/2}$ and $e^{j2\pi}$ are shown to illustrate some examples of this.

where $\Re\{z\} = a$ and $\Im\{z\} = b$ for $z = a + jb$ (compare figs. A.2 and A.3).

This reveals the number's magnitude (absolute value) $|z| = A$ and phase (angle) $\arg(z) = \phi$, which can be extracted by

$$A = |z| = \sqrt{a^2 + b^2} \tag{A.31}$$

$$\phi = \arg(z) = \arctan\left(\frac{b}{a}\right) \tag{A.32}$$

In contrast with this, the $a + jb$ representation is called the *rectangular* form. The algebraic expression that connects the two forms is the complex exponential:

$$z = Ae^{j\phi} = A\cos(\phi) + jA\sin(\phi) \tag{A.33}$$

which follows from Euler's formula,

$$e^{jx} = \cos(x) + j\sin(x) \tag{A.34}$$

This number (e^{jx}) will always lie on the unit circle in the complex plane (Fig. A.3).

The polar form is very useful in signal processing, as it splits complex quantities neatly into amplitudes and phases, which are, in many cases, more meaningful. The complex exponential function $x(t) = e^{j2\pi ft}$ is the fundamental signal for analysis and synthesis, a complex sinusoid that packs a cosine and a sine wave together into a single unit.

The complex exponential form also simplifies some of the arithmetic when it comes to multiplication and division:

$$z \times w = A_z e^{j\phi_z} A_w e^{j\phi_w} = A_z A_w e^{j(\phi_z + \phi_w)} \tag{A.35}$$

$$\frac{z}{w} = \frac{A_z e^{j\phi_z}}{A_w e^{j\phi_w}} = \frac{A_z}{A_w} e^{j(\phi_z - \phi_w)} \tag{A.36}$$

Thus, to multiply we take the product of the magnitudes and the sum of the phases, and to divide we take the quotient of the magnitudes and the difference of the phases. This allows us to reinterpret j as a *rotation* operator that shifts the phase of a complex number by $\pi/2$ radians (90 degrees anticlockwise). Consider the number $z = e^{j\phi} = \cos(\phi) + j\sin(\phi)$, then

$$jz = e^{j(\phi + \pi/2)} = \cos\left(\phi + \frac{\pi}{2}\right) + j\sin\left(\phi + \frac{\pi}{2}\right) = -\sin(\phi) + j\cos(\phi) \tag{A.37}$$

This also shows that $j^2 = -1$, which is equivalent to a shift of π radians.

Finally, we should define the complex conjugate of a number as

$$z = a + jb \quad \Leftrightarrow \quad \bar{z} = a - jb \tag{A.38}$$

that is, the imaginary part changes sign when two numbers are complex conjugates of each other (as shown in Fig. A.3). In polar (complex exponential) form, the magnitudes remain the same, but the phases change sign:

$$z = Ae^{j\phi} \quad \Leftrightarrow \quad \bar{z} = Ae^{-j\phi} \tag{A.39}$$

Some identities apply:

$$
\begin{aligned}
z \times \bar{z} &= a^2 + b^2 = |z|^2 = A^2 \\
z \times \bar{z}^{-1} &= e^{2j\phi} = \frac{z}{|z|} e^{j\phi^*} \\
z + \bar{z} &= 2a = 2\Re\{z\} \\
z - \bar{z} &= 2jb = 2j\Im\{z\}
\end{aligned}
\tag{A.40}
$$

A.4 Complex Polynomials

A complex polynomial function is defined as before (Sect. A.1.3), but using a complex variable instead of a real-valued one:

$$f(z) = \sum_{n=0}^{N} a_n z^n = a_0 + a_1 z + a_2 z^2 + a_3 z^3 + \ldots + a_N z^N \tag{A.41}$$

One such function is called the *z-transform*, which we can use to obtain the *transfer function* of a digital filter whose impulse response $h(t)$ is given:

$$H(z) = \sum_{t=-\infty}^{\infty} h(t) z^{-t} \tag{A.42}$$

where the coefficients of the polynomial are the samples of the impulse response $h(t)$.

For a finite $h(t)$ of size N, Eq. A.42 can be use to give us the filter *frequency response* sampled at points $z = e^{j2\pi k/N}$ (representing sinusoid frequencies $f(k) = 2\pi k/N$ Hz):

$$H(e^{j2\pi k/N}) = \sum_{t=0}^{N-1} h(t) e^{-2\pi kt/N} = DFT(h(t)) \tag{A.43}$$

and so the amplitude and phase responses of this filter at a point k are $|H(e^{j\omega})|$ and $\arg(H(e^{j\omega}))$, respectively, with $\omega = 2\pi k/N$.

In addition, the z-transform can also be used more generally to derive transfer functions and frequency responses of arbitrary linear time-invariant digital filters. For this, the following relationship applies:

$$Y(z) = H(z)X(z) \tag{A.44}$$

where $X(z)$ and $Y(z)$ are the z-transforms of the input and output, respectively, of a filter. $H(z)$, in this case, plays the part of the transfer function. When this is sampled at the DFT frequencies, as in Eq. A.43, we can use it to multiply another spectrum ($X(z)$ sampled at the same frequencies) to accomplish the filtering process (as discussed in Sect. 7.1.4).

The z-transform can be applied directly to the study of filters. For instance the simple filter $y(t) = 0.5 * x(t) + 0.5 * x(t-1)$ $(h(t) = [0.5, 0.5])$, has the following z-transform:

$$H(z) = \sum_{t=0}^{1} h(t) z^{-t} = 0.5 + 0.5 z^{-1} \tag{A.45}$$

From this we can infer a meaning for z^t as a time shift operator, applying a shift of t samples to a signal. So z^{-1} implies a 1-sample delay, z^{-2} implies a 2-sample delay etc. Associated with each delay is a coefficient, which is the value of the impulse response at time t. To get the amplitude and phase responses of this filter, we can evaluate $|H(z)|$ and $\arg(H(z))$ directly for any frequency ω by setting $z = e^{j\omega}$

$$|H(e^{j\omega})| = \sqrt{(0.5+0.5\cos(\omega))^2 + (0.5\sin(\omega))^2} = \cos(\omega/2)$$

$$\arg(H(e^{j\omega})) = \arctan\left(\frac{\sin(\omega)}{0.5+0.5\cos(\omega)}\right) = \frac{\omega}{2} \tag{A.46}$$

which makes this a low pass filter (set $\omega = 2\pi f/f_s$ for any f in Hz to check), with a linear phase shift of $\omega/2$ radians across the frequency spectrum (equivalent to a 1/2-sample delay for the whole signal).

The case of Eq. A.45 is that of an finite impulse response (FIR) filter. For an infinite impulse response (IIR) $y(t) = a_0 x(t) + a_1 x(t-1) + b_1 y(t-1)$, we can first rearrange the equation as

$$y(t) - b_1 y(t-1) = a_0 x(t) + a_1 x(t-1) \tag{A.47}$$

where each side now has its own impulse response $g(t) = [1, -b_1]$ and $f(t) = [a_0, b_0]$. If the z-transforms of $g(t)$ and $f(t)$ are $G(z)$ and $F(z)$, and the filter transfer function is $H(z) = Y(z)/X(z)$ (Eq. A.44), we have

$$Y(z)G(z) = X(z)F(z) \tag{A.48}$$

Therefore, the z-transform $H(z)$ of this filter is the ratio of the z-transforms $G(z)$ and $F(z)$ of $g(t)$ and $f(t)$, respectively:

$$H(z) = \frac{F(z)}{G(z)} = \frac{a_0 + a_1 z^{-1}}{1 - b_1 z^{-1}} \tag{A.49}$$

Thus, for an arbitrary digital filter

$$y(t) = \sum_{n=0}^{N} a_n x(t-n) + \sum_{m=1}^{M} b_m y(t-m) \tag{A.50}$$

the z-transform gives us the transfer function:

$$H(z) = \frac{\sum_{n=0}^{N} a_n z^{-n}}{1 - \sum_{m=1}^{M} b_m z^{-m}} = \frac{a_0 + a_1 z^{-1} + a_2 z^{-2} + ... + a_N z^{-N}}{1 - b_1 z^{-1} - b_2 z^{-2} - ... - b_M z^{-M}} \tag{A.51}$$

Two final remarks: (i) the infinite part of the impulse response is represented by the denominator, and the finite part by the numerator; and (ii) the filter will be stable if the denominator of $|H(z)|$ is not zero, and therefore an FIR filter is always stable.

A.5 Differentiation and Integration

In the real continuum, as the infinitesimal difference between two numbers x and $x + dx$[1] approaches 0, the derivative of a function $f(x)$ can be defined as

$$\frac{df(x)}{dx} \qquad (A.52)$$

This relation is used to get the rate of change of a function and the process of obtaining its value is called *differentiation*. For instance, the derivative of the phase of a sinusoidal signal (as a function of time $\phi(t)$) gives us the instantaneous frequency $f(t)$:

$$f(t) = \frac{d\phi(t)}{dt} \qquad (A.53)$$

The basic rules of differentiation are:

1. For a constant C

$$\frac{dC}{dx} = 0 \qquad (A.54)$$

2. A polynomial term x^n has a derivative

$$\frac{d(x^n)}{dx} = nx^{n-1} \qquad (A.55)$$

3. The derivative of the sum or difference of two terms f and g is

$$\frac{d(f \pm g)}{dx} = \frac{df}{dx} \pm \frac{dg}{dx} \qquad (A.56)$$

4. The derivative of the product of f and g is

$$\frac{d(f \times g)}{dx} = g\frac{df}{dx} + f\frac{dg}{dx} \qquad (A.57)$$

5. The chain rule can be used to work out the derivative of a function of a function (e.g. $df(g(x))/dx$). In this case, we set $y = f(z)$ and $z = g(x)$ in

$$\frac{dy}{dx} = \frac{dy}{dz} \times \frac{dz}{dx}. \qquad (A.58)$$

For trigonometric functions, we have well-known derivatives, such as

$$\frac{d}{dx}\cos(x) = -\sin(x)$$
$$\frac{d}{dx}\sin(x) = \cos(x) \qquad (A.59)$$

---

[1] In this book, the symbol d is sometimes replaced by ∂ (normally used in partial differentiation) to avoid ambiguity if a variable d is already being used.

With these principles, we can work out, for example, the instantaneous frequency of an oscillator whose phase is modulated by a cosine wave, $\sin(\omega t + \cos(\omega t))$:

$$
\begin{aligned}
f(t) = \frac{d\phi(t)}{dt} &= \frac{d(\omega t + \cos(\omega t))}{dt} \\
&= \frac{d\omega t}{dt} + \frac{\cos(\omega t)}{dt} \\
&= \omega + \frac{d\omega t}{dt}\left[\frac{\cos(\omega t)}{d\omega t}\right] \\
&= \omega - \omega\sin(\omega t)
\end{aligned}
\tag{A.60}
$$

The reverse operation to differentiation, in a general sense, is integration. This can take the form of a *definite integral*, over a set interval, or an *indefinite integral*, also known as an *antiderivative*, which is a function $F(x)$ with a derivative $f(x) = dF(x)/dx$. In the former case, the operation gives us a value for the integral, which is a sum over a continuous interval; in the latter case, we have an expression.

In some cases, we can use the rules for differentiation in reverse, as well as some other integration techniques, to work out the antiderivative, which will in turn allow us to calculate a definite integral. However, it is not always possible to guarantee that we will be able to find an antiderivative, and the process is more involved than in the case of differentiation.

The antiderivative of a polynomial term can be worked out using the relation

$$
\int x^n dx = \frac{x^{n+1}}{(n+1)} + C \quad (n \neq -1)
\tag{A.61}
$$

where C is an arbitrary value called the *constant of integration*. Likewise, the antiderivatives of trigonometric functions can be taken from their derivatives:

$$
\begin{aligned}
\int \cos(ax)dx &= \frac{1}{a}\sin(ax) + C \\
\int \sin(ax)dx &= -\frac{1}{a}\cos(ax) + C
\end{aligned}
\tag{A.62}
$$

For example, to find the phase of an oscillator whose frequency is $\omega + \omega\cos(\omega t)$ (FM), we have

$$
\begin{aligned}
\phi(t) = \int \omega + \omega\cos(\omega t)dt \\
= \int \omega dt + \int \omega\cos(\omega t)dt \\
= \omega t + \sin(\omega t) + C
\end{aligned}
\tag{A.63}
$$

So when we modulate the frequency with a cos(), the equivalent PM expression employs a sin(). The meaning of the C constant is that there may be an initial phase offset that we are not able to define when calculating the antiderivative.

Finally, definite integrals are used when we need to realise the sum to obtain a numerical result. This is the case, for instance, for the integral in the Fourier transform formula. Thus is a definite integral, even though the interval goes from $-\infty$ to ∞, because we are using it to calculate a complex spectral coefficient for a given frequency:

$$X(\omega) = \int_{-\infty}^{\infty} x(t)e^{-j\omega t}\,dt \tag{A.64}$$

The dt and the $\int$ symbol go together to represent the fact that this sum is over a continuum (rather than over a discrete series of numbers as in the case of the symbol $\sum$). In practical applications, we can do the integration over a finite interval, such as one cycle of a waveform. In that case we can use the formula:

$$\int_{a}^{b} f(x)dx = F(b) - F(a) \tag{A.65}$$

where $F(x)$ is the antiderivative of $f(x)$. For this to work, we need to find this function, which is more difficult in some cases than in others. However, there are several techniques that can help us get there.

For example, to get the Fourier series coefficients for a square wave,

$$x(t) = \begin{cases} 1 & 0 \le t < \pi \\ -1 & \pi \le t < 2\pi \end{cases} \tag{A.66}$$

we can use

$$X(k) = a_k + jb_k = \int_{0}^{2\pi} x(t)e^{-jkt}\,dt \tag{A.67}$$

The a_k and b_k coefficients are

$$a_k = \int_{0}^{2\pi} x(t)\cos(kt)dt = \frac{1}{2}\left[\int_{0}^{\pi}\cos(kt)dt + \int_{\pi}^{2\pi}-\cos(kt)dt\right]$$
$$= \frac{1}{2}\left[\frac{1}{k}(\sin(0) - \sin(\pi k)) - \frac{1}{k}(\sin(\pi k) - \sin(2\pi k))\right] = 0 \tag{A.68}$$

$$b_k = -\int_{0}^{2\pi} x(t)\sin(kt)dt = -\frac{1}{2}\left[\int_{0}^{\pi}\sin(kt)dt + \int_{\pi}^{2\pi}-\sin(kt)dt\right]$$
$$= -\frac{1}{2}\left[\frac{1}{k}(1 - \cos(\pi k)) - \frac{1}{k}(1(\cos(\pi k) - 1))\right] \tag{A.69}$$
$$= \frac{\cos(\pi k) - 1}{k}$$

Since $\cos(\pi k)$ is -1 for k odd and 1 for k even, the square wave Fourier series (Eq. 7.6) becomes

$$x(t) = 4 \sum_{k=0}^{\infty} \frac{1}{(2k+1)} \sin(2\pi[2k+1]t) \tag{A.70}$$

We can use similar approaches to derive the Fourier series for various waveforms, such as sawtooth (see Eq. 7.10), triangle and pulse waveforms. With some luck, we can find Fourier series for others too.

Appendix B
Application Code

This appendix presents complete applications and other programming examples for the topics studied in this book. Python applications are written for version 3 and depend on the installation of pylab (SciPy, NumPy and matploltlib), as well as Csound and ctcsound.

B.1 Shapes

The full code for the Shapes application discussed in chapter 2 is presented in the listing below:

Listing B.1 Shapes application code.

```
import tkinter
import ctcsound

code = '''
sr=44100
ksmps=64
nchnls=1
0dbfs=1


gar init 0
gifbam2 faustcompile {{
beta = hslider("beta", 0, 0, 2, 0.001);
fbam2(b) = *~(((\(x).(x + x'))*b)+1);
process = fbam2(beta);
}}, "-vec -lv 1"

gar init 0
```

```
instr 1
ival = p4
kp chnget "pitch"
kp tonek kp, 10, 1
kv chnget "volume"
kv tonek kv, 10
ain oscili 1, p5*kp, -1, 0.25
ib, asig faustaudio gifbam2,ain
kb1 expsegr ival,0.01,ival,30,0.001,0.2,0.001
faustctl ib,"beta",kb1
asig balance asig, ain
kenv expsegr 1,20,0.001,0.2,0.001
aenv2 linsegr 0,0.015,0,0.001,p4,0.2,p4
aout = asig*aenv2*kenv*0.4*kv
 out  aout
gar += aout
endin

instr 100

a1,a2 reverbsc gar,gar,0.7,2000
amix = (a1+a2)/2
out amix
ks rms gar+amix
chnset ks, "meter"
gar = 0
endin

schedule(100,0,-1)
'''


class Application(tkinter.Frame):
 def move(self,event):
    canvas = event.widget
    x = canvas.canvasx(event.x)
    y = canvas.canvasy(event.y)
    item = canvas.find_withtag("current")[0]
    canvas.coords(item, x+10, y+10, x-10, y-10)
    self.cs.setControlChannel("pitch",
        0.5+1.5*x/self.size)
    self.cs.setControlChannel("volume",
        2.0*(self.size-y)/self.size)

 def play(self,event):
    note = event.widget.find_withtag("current")[0]
```

```
        self.canvas.itemconfigure(note, fill="red")
        self.perf.inputMessage("i1 0 -1 0.5 440")

    def stop(self,event):
        note = event.widget.find_withtag("current")[0]
        self.canvas.itemconfigure(note, fill="black")
        self.perf.inputMessage("i-1 0 0.5 440")

    def createCanvas(self):
        self.size = 600
        self.canvas = tkinter.Canvas(self,height=self.size,
                            width=self.size, bg="violet")
        self.canvas.pack()

    def createCircle(self):
        circle = self.canvas.create_oval(self.size/2-10,
                                    self.size/2-10,
                                    self.size/2+10,
                                    self.size/2+10,
                                    fill="black")

        self.canvas.tag_bind(circle,"<ButtonPress>",
                        self.play)
        self.canvas.tag_bind(circle,"<B1-Motion>",
                        self.move)
        self.canvas.tag_bind(circle,"<ButtonRelease>",
                        self.stop)

    def createMeter(self):
     iw = 10
     self.vu = []
     for i in range(0, self.size, iw):
      self.vu.append(self.canvas.create_rectangle(i,
          self.size-40,i+iw,self.size,fill="grey"))

    def drawMeter(self):
        level,err = self.cs.controlChannel("meter")
        cnt = 0
        level *= 16000
        red = (self.size/10)*0.8
        yellow = (self.size/10)*0.6
        for i in self.vu:
          if level > cnt*100:
            if cnt > red:
             self.canvas.itemconfigure(i, fill="red")
```

```
      elif cnt > yellow:
        self.canvas.itemconfigure(i, fill="yellow")
      else:
        self.canvas.itemconfigure(i, fill="blue")
    else:
      self.canvas.itemconfigure(i, fill="grey")
    cnt  = cnt + 1
  self.master.after(50,self.drawMeter)

def quit(self):
  self.perf.stop()
  self.perf.join()
  self.master.destroy()

def createEngine(self):
  self.cs = ctcsound.Csound()
  res = self.cs.compileOrc(code)
  self.cs.setOption('-odac')
  if res == 0:
    self.cs.start()
    self.cs.setControlChannel('pitch', 1.0)
    self.cs.setControlChannel('volume', 1.0)
    self.perf =
      ctcsound.CsoundPerformanceThread(self.cs.csound())
    self.perf.play()
    return True
  else:
    return False

def __init__(self,master=None):
  tkinter.Frame.__init__(self,master)
  self.master.title('Csound + Tkinter: '
    'just click and play')
  self.master = master
  self.pack()
  self.createCanvas()
  self.createCircle()
  self.createMeter()
  if self.createEngine() is True:
    self.drawMeter()
    self.master.protocol('WM_DELETE_WINDOW',
                         self.quit)
    self.master.mainloop()
  else: self.master.quit()
```

```
Application(tkinter.Tk())
```

B.2 Vocal Quartet Simulation

Listing B.2 provides a full example of a vocal quartet simulation, based on a variant of the instrument discussed in chapter 3, including the use of a global reverb instrument.

Listing B.2 Csound vocal simulation.

```
gasig init 0
opcode Filter,a,akk
 as,kf,kbw xin
 as = reson(as,kf,kbw,1)
 as = reson(as,kf,kbw,1)
    xout(as)
endop

instr 1
 kf[][] init 5,5
 kb[][] init 5,5
 ka[][] init 5,5

 kf[][] fillarray 800,1150,2800,3700,4950,
                  450,800,2830,3500,4950,
                  400,1600,2700,3500,4950,
                  350,1700,2700,3700,4950,
                  325,700,2530,3500,4950
 kb[][] fillarray 80,90,120,130,140,
                  70,80,100,130,135,
                  60,80,120,150,200,
                  50,100,120,150,200,
                  50,60,170,180,200
 ka[][] fillarray 0,4,20,36,60,
                  0,9,16,28,55,
                  0,24,30,35,60,
                  0,20,30,36,60,
                  0,12,30,40,64

 kv = linseg(p6,p3/2-p7,p6,p7,p8,p3/2,p8)
 itim = p7

 kf0 = port(kf[kv][0],itim,i(kf[p6][0]))
```

```
  kf1 = port(kf[kv][1],itim,i(kf[p6][1]))
  kf2 = port(kf[kv][2],itim,i(kf[p6][2]))
  kf3 = port(kf[kv][3],itim,i(kf[p6][3]))
  kf4 = port(kf[kv][4],itim,i(kf[p6][4]))

  kb0 = port(kb[kv][0],itim,i(kb[p6][0]))
  kb1 = port(kb[kv][1],itim,i(kb[p6][1]))
  kb2 = port(kb[kv][2],itim,i(kb[p6][2]))
  kb3 = port(kb[kv][3],itim,i(kb[p6][3]))
  kb4 = port(kb[kv][4],itim,i(kb[p6][4]))

  ka0 = port(ampdb(-ka[kv][0]),itim,-i(ka[p6][0]))
  ka1 = port(ampdb(-ka[kv][1]),itim,-i(ka[p6][0]))
  ka2 = port(ampdb(-ka[kv][2]),itim,-i(ka[p6][0]))
  ka3 = port(ampdb(-ka[kv][3]),itim,-i(ka[p6][0]))
  ka4 = port(ampdb(-ka[kv][4]),itim,-i(ka[p6][0]))

  kjit = randi(p5*0.01,15,2)
  kvib = oscili(p5*(0.01+
            gauss:k(0.005)),
            4+gauss:k(0.05))

  kfun = p5+kjit+kvib
  a1 = buzz(p4,kfun,sr/(2*kfun),-1,rnd(0.1))
  kf0 = kfun > kf0 ? kfun : kf0
  af1 = Filter(a1*ka0,kf0,kb0)
  af2 = Filter(a1*ka1,kf1,kb1)
  af3 = Filter(a1*ka2,kf2,kb2)
  af4 = Filter(a1*ka3,kf3,kb3)
  af5 = Filter(a1*ka4,kf4,kb4)

  amix = delay(af1+af2+af3+af4+af5,rnd(0.1))
  asig= linenr(amix*5,0.01,0.05,0.01)
    out(asig)
    gasig += asig
endin

instr 2
 a1,a2 reverbsc gasig,gasig,0.8,3000
  out (a1+a2)/3
  gasig = 0
endin

schedule(2,0,-1)
icnt init 0
```

```
while icnt < 4 do
schedule(1,0+icnt*20,4.9,0dbfs,cpspch(7.04),
        3,1+gauss(0.3),4)
schedule(1,5+icnt*20,4.9,0dbfs,cpspch(7.02),
        4,1+gauss(0.3),2)
schedule(1,10+icnt*20,4.9,0dbfs,cpspch(7.00),
        2,1+gauss(0.3),1)
schedule(1,15+icnt*20,4.9,0dbfs,cpspch(6.11),
        1,1+gauss(0.3),3)
if icnt > 0 then
 schedule(1,0+icnt*20,4.9,0dbfs,cpspch(8.07),
        3,1+gauss(0.3),4)
 schedule(1,5+icnt*20,4.9,0dbfs,cpspch(8.06),
        4,1+gauss(0.3),2)
 schedule(1,10+icnt*20,4.9,0dbfs,cpspch(8.04),
        2,1+gauss(0.3),1)
 schedule(1,15+icnt*20,4.9,0dbfs,cpspch(8.06),
        1,1+gauss(0.3),3)
endif
if icnt > 1 then
 schedule(1,0+icnt*20,4.9,0dbfs*1.5,cpspch(9.04),
        3,1+gauss(0.3),4)
 schedule(1,5+icnt*20,4.9,0dbfs*1.2,cpspch(8.11),
        4,1+gauss(0.3),2)
 schedule(1,10+icnt*20,4.9,0dbfs,cpspch(9.00),
        2,1+gauss(0.3),1)
 schedule(1,15+icnt*20,4.9,0dbfs*1.3,cpspch(9.03),
        1,1+gauss(0.3),3)
endif
if icnt > 2 then
 schedule(1,0+icnt*20,4.9,0dbfs,cpspch(7.07),
        3,1+gauss(0.3),4)
 schedule(1,5+icnt*20,4.9,0dbfs,cpspch(7.11),
        4,1+gauss(0.3),2)
 schedule(1,10+icnt*20,4.9,0dbfs,cpspch(7.07),
        2,1+gauss(0.3),1)
 schedule(1,15+icnt*20,4.9,0dbfs,cpspch(7.09),
        1,1+gauss(0.3),3)
 endif
icnt += 1
od

schedule(1,icnt*20,4.9,0dbfs,cpspch(7.04),
        3,1+gauss(0.3),4)
schedule(1,icnt*20,4.9,0dbfs,cpspch(8.07),
```

```
          3,1+gauss(0.3),4)
schedule(1,icnt*20,4.9,0dbfs,cpspch(9.04),
          3,1+gauss(0.3),4)
schedule(1,icnt*20,4.9,0dbfs,cpspch(7.11),
          3,1+gauss(0.3),4)
event_i("e",0,icnt*20+8,0)
```

B.3 Closed-Form Summation Formulae User-Defined Opcodes

The following listings provide ready-to-use implementations of the techniques explored in chapter 4.

Listing B.3 Bandlimited pulse UDO.

```
/********************************
asig Blp kamp,kfreq
kamp - amplitude
kfreq - fundamental frequency
*********************************/
opcode Blp,a,kki
  setksmps 1
  kamp,kf  xin
  kn = int(sr/(2*kf))
  kph phasor kf/2
  kden tablei kph,-1,1
  if kden != 0 then
      knum tablei kph*(2*kn+1),-1,1,0,1
      asig = (kamp/(2*kn))*(knum/kden - 1)
   else
    asig = kamp
  endif
  xout asig
endop
```

Listing B.4 Generalised summation-formulae UDO.

```
/********************************
asig Blsum kamp,kfr1,kfr2,ka
kamp - amplitude
kfr1 - frequency 1 (omega)
kfr2 - frequency 2 (theta)
ka - distortion amount
*********************************/
opcode Blsum,a,kkkki
```

```
kamp,kw,kt,ka xin
kn = int(((sr/2) - kw)/kt)
aphw phasor kw
apht phasor kt
a1 tablei aphw,-1,1
a2 tablei aphw - apht,-1,1,0,1
a3 tablei aphw + (kn+1)*apht,-1,1,0,1
a4 tablei aphw + kn*apht,-1,1,0,1
acos tablei apht,-1,1,0.25,1
kpw pow ka,kn+1
ksq = ka*ka
aden = (1 - 2*ka*acos + ksq)
asig = (a1 - ka*a2 - kpw*(a3 - ka*a4))/aden
knorm = sqrt((1-ksq)/(1 - kpw*kpw))
xout asig*kamp*knorm
endop
```

Listing B.5 Non-bandlimited generalised summation-formulae UDO.

```
/********************************
asig NBlsum kamp,kfr1,kfr2,ka
kamp - amplitude
kfr1 - frequency 1 (omega)
kfr2 - frequency 2 (theta)
ka - distortion amount
*********************************/
opcode NBlsum,a,kkkk
 kamp,kw,kt,ka   xin
 aphw phasor kw
 apht phasor kt
 a1 tablei aphw,-1,1
 a2 tablei aphw - apht,itb,1,0,1
 acos tablei apht,-1,1,0.25,1
 ksq = ka*ka
 asig = (a1 - ka*a2)/(1 - 2*ka*acos + ksq)
 knorm = sqrt(1-ksq)
 xout asig*kamp*knorm
endop
```

Listing B.6 PM synthesis UDO.

```
/**************************************************
asig  PM kamp,kfc,kfm,kndx
kamp - amplitude
kfc - carrier frequency
kfm - modulation frequency
kndx - distortion index
```

```
*************************************************/
opcode PM,a,kkkk
 kamp,kfc,kfm,kndx xin
 acar phasor kfc
 amod oscili kndx/(2*$M_PI),kfm
 apm  tablei acar+amod,-1,1,0.25,1
     xout apm*kamp
endop
```

Listing B.7 PM operator UDO.

```
/*********************************************
asig PMOp kamp,kfr,apm,iatt,idec,isus,irel[,ifn]
kamp - amplitude
kfr - frequency
apm - phase modulation input
iatt - attack
idec - decay
isus - sustain
irel - release
ifn - optional wave function table (defaults to sine)
***********************************************/
opcode PMOp,a,kkaiiiij
 kmp,kfr,apm,
    iatt,idec,
    isus,irel,ifn xin
 aph phasor kfr
 a1  tablei aph+apm/(2*$M_PI),ifn,1,0,1
 a2  madsr iatt,idec,isus,irel
     xout  a2*a1*kmp
endop
```

Listing B.8 Asymmetric FM UDO.

```
f5 0 131072 "exp" 0 -50 1
/*********************************************
asig Asfm kamp,kfc,kfm,kndx,kR,ifn,imax
kamp - amplitude
kfc - carrier frequency
kfm - modulation frequency
kndx - distortion index
ifn - exp func between 0 and -imax
imax - max absolute value of exp function
***********************************************/
opcode Asfm,a,kkkkkii
 kamp,kfc,kfm,knx,kR,ifn,imax
 kndx = knx*(kR+1/kR)*0.5
```

```
kndx2 = knx*(kR-1/kR)*0.5
afm oscili kndx/(2*$M_PI),kfm
aph phasor kfc
afc tablei aph+afm,ifn,1,0,1
amod oscili kndx2, kfm, -1, 0.25
aexp tablei -(amod-abs(kndx2))/imx, ifn, 1
         xout kamp*afc*aexp
endop
```

Listing B.9 PAF UDO.

```
opcode Func,a,a
 asig xin
        xout 1/(1+asig^2)
endop


/***************************************
asig PAF kamp,kfun,kcf,kfshift,kbw
kamp - amplitude
kfun - fundamental freq
kcf - centre freq
kfshift - shift freq
kbw - bandwidth
***************************************/
opcode PAF,a,kkkkki
  kamp,kfo,kfc,kfsh,kbw   xin
  kn = int(kfc/kfo)
  ka = (kfc - kfsh - kn*kfo)/kfo
  kg = exp(-kfo/kbw)
  afsh phasor kfsh
  aphs phasor kfo/2
  a1 tablei 2*aphs*kn+afsh,-1,1,0.25,1
  a2 tablei 2*aphs*(kn+1)+afsh,-1,1,0.25,1
  asin tablei aphs, 1, 1, 0, 1
  amod Func 2*sqrt(kg)*asin/(1-kg)
  kscl = (1+kg)/(1-kg)
  acar = ka*a2+(1-ka)*a1
  asig = kscl*amod*acar
          xout asig*kamp
endop
```

Listing B.10 ModFM UDO.

```
/*********************************************
asig ModFM kamp,kfc,kfm,kndx,ifn,imax
kamp - amplitude
kfc - carrier frequency
```

```
kfm - modulation frequency
kndx - distortion index
ifn - exp func between 0 and -imax
imax - max  absolute value of exp function
*************************************************/
opcode ModFM,a,kkkkiii
 kamp,kfc,kfm,kndx,iexp,imx xin
 acar oscili kamp,kfc,-1,0.25
 acos oscili 1,kfm,-1,0.25
 amod table -kndx*(acos-1)/imx,iexp,1
          xout acar*amod
endop
```

Listing B.11 ModFM formant synthesis UDO.

```
/*********************************************
asig ModForm kamp,kfo,kfc,kbw,ifn,imax
kamp - amplitude
kfo - fundamental frequency
kfc - formant centre frequency
kbw - bandwidth
ifn - exp func between 0 and -imax
imax - max absolute value of exp function
*************************************************/
opcode ModForm,a,kkkkii
 kamp,kfo,kfc,kbw,ifn,itm  xin
 ioff = 0.25
 itab = -1
 icor = 4.*exp(-1)
 ktrig changed kbw
 if ktrig == 1 then
  k2 = exp(-kfo/(.29*kbw*icor))
  kg2 = 2*sqrt(k2)/(1.-k2)
  kndx = kg2*kg2/2.
 endif
 kf = kfc/kfo
 kfin = int(kf)
 ka = kf  - kfin
 aph    phasor kfo
 acos tablei aph, 1, 1, 0.25, 1
 aexp table kndx*(1-acos)/itm,ifn,1
 acos1 tablei aph*kfin, itab, 1, ioff, 1
 acos2 tablei aph*(kfin+1), itab, 1, ioff, 1
 asig = (ka*acos2 + (1-ka)*acos1)*aexp
 xout asig*kamp
endop
```

Listing B.12 Waveshaping UDO.

```
/************************************************
asig  Waveshape kamp,kfreq,kndx,ifn1,ifn2
kamp - amplitude
kfreq - frequency
kndx - distortion index
ifn1 - transfer function
ifn2 - scaling function
*************************************************/
opcode Waveshape,a,kkkiii
 kamp,kf,kndx,itf,igf xin
 asin oscili 0.5*kndx,kf
 awsh tablei asin,itf,1,0.5
 kscl tablei kndx,igf,1
 xout awsh*kamp*kscl
endop
```

Listing B.13 Sawtooth wave oscillator based on waveshaping.

```
f2 0 16385 "tanh" -157 157
f3 0 8193 4 2 1
/************************************************
asig  Sawtooth kamp,kfreq,kndx,ifn1,ifn2
kamp - amplitude
kfreq - frequency
kndx - distortion index
ifn1 - transfer function
ifn2 - scaling function
*************************************************/
opcode Sawtooth,a,kkkii
 kamp,kf,kndx,itf,igf xin
 amod oscili 1,kf,-1,0.25
 asq Waveshape kamp*0.5,kf,kndx,-1,itf,igf
         xout asq*(amod + 1)
endop
```

B.4 Pylab Waveform and Spectrum Plots

The code for plotting waveforms and spectra from an input signal is presented in
Listing B.14. The function in this example can be dropped into any Python program
that produces a signal in a NumPy array.

Listing B.14 Pylab example code for waveform and spectrum plotting.

```
import pylab as pl

def plot_signal(s,sr,start=0,end=440,N=32768):
  '''plots the waveform and spectrum of a signal s
     with sampling rate sr, from a start to an
     end position (waveform), and from a start
     position with a N-point DFT (spectrum)'''
  pl.figure(figsize=(8,5))

  pl.subplot(211)
  sig = s[start:end]
  time = pl.arange(0,len(sig))/sr
  pl.plot(time,sig, 'k-')
  pl.ylim(-1.1,1.1)
  pl.xlabel("time (s)")

  pl.subplot(212)
  N = 32768
  win = pl.hanning(N)
  scal = N*pl.sqrt(pl.mean(win**2))
  sig = s[start:start+N]
  window = pl.rfft(sig*win/max(sig))
  f = pl.arange(0,len(window))
  bins = f*sr/N
  mags = abs(window/scal)
  spec = 20*pl.log10(mags/max(mags))
  pl.plot(bins,spec, 'k-')
  pl.ylim(-60, 1)
  pl.ylabel("amp (dB)", size=16)
  pl.xlabel("freq (Hz)", size=16)
  pl.yticks()
  pl.xticks()
  pl.xlim(0,sr/2)

  pl.tight_layout()
  pl.show()
```

B.5 FOF Vowel Synthesis

The code in Listing B.15 demonstrates FOF synthesis of vowel sounds, together with a transition to a granular texture. The instrument has a large number of parameters, defining the time-varying features of the sound:

- p4: amplitude
- p5, p6: start, end fundamental
- p7 - p11: formant amplitudes
- p12, p13: start, end octave transposition
- p14, p15: start, end grain attack time
- p16, p17: start, end bandwidth
- p18, p19: start, end grain size
- p20, p21: start, end vowel cycle frequency
- p22: vowel jitter

Listing B.15 FOF vowel synthesis and granulation.

```
gif = ftgen(0,0,16384,7,0,16384,1)
gif1 = ftgen(0,0,8,-2,400,604,325,325,
                360,604,238,360)
gif2 = ftgen(0,0,8,-2,1700,1000,700,700,
                750,1000,1741,750)
gif3 = ftgen(0,0,8,-2,2300,2450,2550,2550,
                2400,2450,2450,2400)
gif4 = ftgen(0,0,8,-2,2900,2700,2850,2850,
                2675,2700,2900,2675)
gif5 = ftgen(0,0,8,-2,3400,3240,3100,3100,
                2950,3240,4000,2950)

instr 1
 iscal = p4/(p7+p8+p9+p10+p11)
 iol = 300

 ksl = line(p20,p3,p21)
 kv = rand(p22)
 kff = phasor(ksl+kv)
 kf1 = tablei(kff,gif1,1,0,1)
 kf2 = tablei(kff,gif2,1,0,1)
 kf3 = tablei(kff,gif3,1,0,1)
 kf4 = tablei(kff,gif4,1,0,1)
 kf5 = tablei(kff,gif5,1,0,1)

 koc = linseg(p12,p3*.4,p12,p3*.3,p13,p3*.3,p13)
 kat = linseg(p14,p3*.3,p14,p3*.2,p15,p3*.5,p15)
```

```
kbw = linseg(p16,p3*.25,p16,p3*.15,p17,p3*.6,p17)
ks  = linseg(p18,p3*.25,p18,p3*.15,p19,p3*.6,p19)
idc = .007

aj1 = randi(.02,2)
aj2 = randi(.01,.75)
aj3 = randi(.02,1.2)
av = oscil(.007,7)
kf = line(p5, p3, p6)
af =(cpspch(kf))*(aj1+aj2+aj3+1)*(av+1)

a1 = fof(p7,af,kf1,koc,kbw,kat,ks,idc,iol,-1,gif,p3)
a2 = fof(p8,af,kf2,koc,kbw,kat,ks,idc,iol,-1,gif,p3)
a3 = fof(p9,af,kf3,koc,kbw,kat,ks,idc,iol,-1,gif,p3)
a4 = fof(p10,af,kf4,koc,kbw,kat,ks,idc,iol,-1,gif,p3)
a5 = fof(p11,af,kf5,koc,kbw,kat,ks,idc,iol,-1,gif,p3)
amx = (a1+a2+a3+a4+a5)*iscal
     out(amx)
endin

schedule(1,0,60,
        0dbfs/3,6.09,6.07,
        1,.5,.3,.1,.01,
        0,3,.003,.01,
        100,10,.02,.1,
        0.5,.01,.01)
```

B.6 Convolution Programs

The following program outputs the convolution of two input files, using an FFT overlap-add method. It works with audio files of the RIFF-Wave type, taking three arguments,

```
$ python3 conv.py input1.wav input2.wav output.wav
```

Listing B.16 FFT-based convolution program.

```
import pylab as pl
import scipy.io.wavfile as wf
import sys

def conv(signal,ir):
```

```
    N = len(ir)
    L = len(signal)
    M = 2
    while(M <= 2*N-1): M *= 2

    h = pl.zeros(M)
    x = pl.zeros(M)
    y = pl.zeros(L+N-1)

    h[0:N] = ir
    H = pl.rfft(h)
    n = N
    for p in range(0,L,N):
     if p+n > L:
          n = L-p
          x[n:] = pl.zeros(M-n)
     x[0:n] = signal[p:p+n]
     y[p:p+2*n-1] += pl.irfft(H*pl.rfft(x))[0:2*n-1]
    return y

(sr,x1) = wf.read(sys.argv[1])
(sr,x2) = wf.read(sys.argv[2])
scal = 32768
if len(x1) > len(x2):
 y = conv(x1,x2/scal)
 a = max(x1)
else:
 y = conv(x2, x1/scal)
 a = max(x2)

s = max(y)
out = pl.array(y*a/s,dtype='int16')
wf.write(sys.argv[3],sr,out)
```

The next program applies the partitioned convolution method to split the impulse response into segments of 1024 samples. It works with the same command-line parameters as in the previous case.

Listing B.17 Uniform partitioned convolution program.

```
iimport pylab as pl
import scipy.io.wavfile as wf
import sys

def pconv(input,ir,S=1024):
    N = len(ir)
    L = len(input)
```

```python
    M = S*2

    P = int(pl.ceil(N/S))
    H = pl.zeros((P,M//2+1)) + 0j
    X = pl.zeros((P,M//2+1)) + 0j
    x = pl.zeros(M)
    o = pl.zeros(M)
    s = pl.zeros(S)

    for i in range(0,P):
        p = (P-1-i)*S
        ll = len(ir[p:p+S])
        x[:ll] = ir[p:p+S]
        H[i] = pl.rfft(x)

    y = pl.zeros(L+N-1)
    n = S
    i = 0
    for p in range(0,L+N-S,S):
      if p > L:
        x = pl.zeros(M)
      else:
       if n+p > L:
         n = L - p
         x[n:] = pl.zeros(M-n)
       x[:n] = input[p:p+n]
      X[i] = pl.rfft(x)
      i = (i+1)%P
      O = pl.zeros(M//2+1) + 0j
      for j in range(0,P):
        k = (j+i)%P
        O += H[j]*X[k]
      o = pl.irfft(O)
      y[p:p+S] = o[:S] + s
      s = o[S:]

    return y

(sr,x1) = wf.read(sys.argv[1])
(sr,x2) = wf.read(sys.argv[2])

scal = 32768
L1 = len(x1)
L2 = len(x2)
y = pl.zeros(L1+L2-1)
```

```
m = 0
if L1 > L2:
 y = pconv(x1,x2/scal)
 a = max(x1)
else:
 y = pconv(x2,x1/scal)
 a = max(x2)
s = max(y)
scal = a/s
out = pl.array(y*scal,dtype='int16')
wf.write(sys.argv[3],sr,out)
```

The next example is the implementation of non-uniform partitioned convolution in Csound, as a UDO. It uses dconv for direct convolution and ftconv for fast partitioned convolution.

Listing B.18 Csound UDO for non-uniform partitioned convolution.

```
/***********************************************
asig  Pconv ain,ifn
kamp - input signal
ifn - IR function table
***********************************************/
opcode Pconv,a,ai
 asig,ifn xin
 a1 dconv asig,32,ifn
 a2 ftconv asig,ifn,32,32,96
 a3 ftconv asig,ifn,128,128,896
 a4 ftconv asig,ifn,1024,1024
 xout a1+a2+a3+a4
endop
```

B.7 Spectral Masking

The following program is a skeleton for a spectral masking application. The function mask(mags, ...) can be replaced by a different user-defined operation on the magnitude spectrum, taking the original values and returning a modified array. The example supplied creates a band-pass filter centred at fc. The command-line for this program takes two file names (input and output):

```
$ python3 mask.py input1.wav output.wav
```

Listing B.19 Spectral masking program.

```python
import pylab as pl
import scipy.io.wavfile as wf
import sys

def stft(x,w):
    X = pl.rfft(w*x)
    return X

def istft(X,w):
    xn = pl.irfft(X)
    return xn*w

def p2r(mags, phs):
    return mags*pl.cos(phs) + 1j*mags*pl.sin(phs)

def mask(mags,fc):
    m = pl.zeros(N//2+1)
    d = int((fc/2)*N/sr)
    b = int(fc*N/sr)
    m[b-d:b+d] = pl.hanning(2*d)
    mags *= m
    return mags

N = 1024
D = 4
H = N//D
zdbfs = 32768
(sr,signal) = wf.read(sys.argv[1])
signal = signal/zdbfs
L = len(signal)
output = pl.zeros(L)
win = pl.hanning(N)
scal = 1.5*D/4

fc = 1000
fe = 5000
incr = (fe - fc)*(H/L)

for n in range(0,L,H):
    if(L-n < N): break
    frame = stft(signal[n:n+N],win,n)
    mags = abs(frame)
    phs = pl.angle(frame)
    mags = mask(mags,fc)
```

```
    fc += incr
    frame = p2r(mags,phs)
    output[n:n+N] += istft(frame,win,n)

output = pl.array(output*zdbfs/scal,dtype='int16')
wf.write(sys.argv[2],sr,output)
```

B.8 Cross-synthesis

In this section we present a cross-synthesis program example. The process implemented here is the spectral vocoder (see 7.2.2), but this code can be used as a skeleton for other types of cross-synthesis applications. The command-line for this program is:

```
$ python3 cross.py input1.wav input2.wav output.wav
```

Listing B.20 Spectral vocoder program.

```python
import pylab as pl
import scipy.io.wavfile as wf
import sys

def stft(x,w):
    X = pl.rfft(w*x)
    return X

def istft(X,w):
    xn = pl.irfft(X)
    return xn*w

def p2r(mags, phs):
    return mags*pl.cos(phs) + 1j*mags*pl.sin(phs)

def spec_env(frame,coefs):
    mags = abs(frame)
    N = len(mags)
    ceps = pl.rfft(pl.log(mags[:N-1]))
    ceps[coefs:] = 0
    mags[:N-1] = pl.exp(pl.irfft(ceps))
    return mags

N = 1024
```

```
D = 4
H = N//D
zdbfs = 32768
(sr,in1) = wf.read(sys.argv[1])
(sr,in2) = wf.read(sys.argv[2])
L1 = len(in1)
L2 = len(in2)
if L2 > L1: L = L2
else: L = L1
signal1 = pl.zeros(L)
signal2 = pl.zeros(L)
signal1[:len(in1)] = in1/zdbfs
signal2[:len(in2)] = in2/zdbfs
output = pl.zeros(L)
win = pl.hanning(N)
scal = 1.5*D/4

for n in range(0,L,H):
    if(L-n < N): break
    frame1 = stft(signal1[n:n+N],win,n)
    frame2 = stft(signal2[n:n+N],win,n)
    mags = abs(frame1)
    env1 = spec_env(frame1,20)
    env2 = spec_env(frame2,20)
    phs = pl.angle(frame1)
    if(min(env1) > 0):
      frame = p2r(mags*env2/env1,phs)
    else:
      frame = p2r(mags*env2,phs)
    output[n:n+N] += istft(frame,win,n)

a = max(signal1)
b = max(output)
c = a/b
output = pl.array(output*zdbfs*c/scal,dtype='int16')
wf.write(sys.argv[3],sr,output)
```

B.9 Pitch Shifting

A pitch shifter with formant preservation is shown in Listing B.21. A version of this program without this feature can be created by removing the spectral envelope scaling factor in the phase vocoder synthesis method call. Its command line takes a number for the pitch ratio and two files:

```
$ python3 pshift.py <ratio> input.wav output.wav
```

Listing B.21 Pitch shifter program with formant preservation.

```python
import pylab as pl
import scipy.io.wavfile as wf
import sys

def stft(x,w,n):
    N = len(w)
    X = pl.rfft(x*w)
    k = pl.arange(0,N/2+1)
    return X*pl.exp(-2*pl.pi*1j*k*n/N)

def istft(X,w,n):
    N = len(w)
    k = pl.arange(0,N/2+1)
    xn = pl.irfft(X*pl.exp(2*pl.pi*1j*k*n/N))
    return xn*w

def modpi(x):
    if x >= pl.pi: x -= 2*pl.pi
    if x < -pl.pi: x += 2*pl.pi
    return x
unwrap = pl.vectorize(modpi)

class Pvoc:
    def __init__(self,w,h,sr):
        N = len(w)
        self.win = w
        self.phs = pl.zeros(N//2+1)
        self.h = h
        self.n = 0
        self.fc = pl.arange(N//2+1)*sr/N
        self.sr = sr

    def analysis(self,x):
        X = stft(x,self.win,self.n)
        delta = pl.angle(X) - self.phs
        self.phs = pl.angle(X)
        delta = unwrap(delta)
        f = self.fc + delta*self.sr/(2*pl.pi*self.h)
        self.n += self.h
        return abs(X),f
```

```python
    def synthesis(self,a,f):
        delta = (f - self.fc)*(2*pl.pi*self.h)/self.sr
        self.phs += delta
        X = a*pl.cos(self.phs) + 1j*a*pl.sin(self.phs)
        y = istft(X,self.win,self.n)
        self.n += self.h
        return y


N = 1024
D = 8
H = N//D
zdbfs = 32768
(sr,signal) = wf.read(sys.argv[2])
signal = signal/zdbfs
L = len(signal)
output = pl.zeros(L)
win = pl.hanning(N)

pva = Pvoc(win,H,sr)
pvs = Pvoc(win,H,sr)
trans = float(sys.argv[1])

def scale(amps,freqs,p):
    N = len(freqs)
    a,f = pl.zeros(N),pl.zeros(N)
    for i in range(0, N):
        n = int((i*p))
        if n > 0 and n < N-1:
            f[n] = p*freqs[i]
            a[n] = amps[i]
    return a,f

def true_spec_env(amps,coefs,thresh):
    N = len(amps)
    sm = 10e-15
    form = pl.zeros(N)
    lmags = pl.log(amps[:N-1]+sm)
    mags = lmags
    check = True
    while(check):
        ceps = pl.rfft(lmags)
        ceps[coefs:] = 0
        form[:N-1] = pl.irfft(ceps)
        for i in range(0,N-1):
```

```
        if lmags[i] < form[i]: lmags[i] = form[i]
        diff = mags[i] - form[i]
        if diff > thresh: check = True
        else: check = False
    return pl.exp(form)+sm

for n in range(0,L,H):
    if(L-n < N): break
    amps,freqs = pva.analysis(signal[n:n+N])
    env1 = true_spec_env(amps,40,0.23)
    amps,freqs = scale(amps,freqs,trans)
    env2 = true_spec_env(amps,40,0.23)
    output[n:n+N] += pvs.synthesis(amps*env1/env2,freqs)

scal = max(output)/max(signal)
output = pl.array(output*zdbfs/scal,dtype='int16')
wf.write(sys.argv[3],sr,output)
```

B.10 Time Scaling

The example in this section implements time scaling with phase locking. Its command line is:

```
$ python3 tscale.py <ratio> input.wav output.wav
```

Listing B.22 Time scaling program

```
import pylab as pl
import scipy.io.wavfile as wf
import sys

def stft(x,w):
    X = pl.rfft(x*w)
    return X

def istft(X,w):
    xn = pl.irfft(X)
    return xn*w

N = 1024
D = 8
H = N//D
```

```
zdbfs = 32768
(sr,signal) = wf.read(sys.argv[2])
signal = signal/zdbfs
L = len(signal)
win = pl.hanning(N)
Z = pl.zeros(N//2+1)+0j+10e-20
np = 0
ts = float(sys.argv[1])
output = pl.zeros(int(L/ts)+N)
ti = int(ts*H)
scal = 3*D/4

def plock(x):
    N = len(x)
    y = pl.zeros(N)+0j
    y[0], y[N-1] = x[0], x[N-1]
    y[1:N-1] = x[1:N-1] - (x[0:N-2] + x[2:N])
    return y

Z = pl.zeros(N//2+1)+0j+10e-20
for n in range(0,L-(N+H),ti):
    X1 = stft(signal[n:n+N],win)
    X2 = stft(signal[n+H:n+H+N],win)
    Y = X2*(Z/X1)*abs(X1/Z)
    Z = plock(Y)
    output[np:np+N] += istft(Y,win)
    np += H

output = pl.array(output*zdbfs/scal,dtype='int16')
wf.write(sys.argv[3],sr,output)
```

References

1. Allan, R.: A History of the Personal Computer: The People and the Technology. Allan Pub. (2001)
2. Arfib, D.: Digital synthesis of complex spectra by means of multiplication of non-linear distorted sine waves. In: Audio Engineering Society Convention 59 (1978)
3. Askill, J.: The Physics of Musical Sound. Van Nostrand, New York (1979)
4. Atmel Corp.: Home page Atmel Microprocessors (2016). URL http://www.atmel.com
5. Backus, J.: The Acoustical Foundations of Music. W.W. Norton, New York (1977)
6. Beauchamp, J.: Analysis and synthesis of musical instrument sounds. In: J. Beauchamp (ed.) Analysis, Synthesis, and Perception of Musical Sounds: The Sound of Music, Modern Acoustics and Signal Processing, pp. 1–89. Springer, New York (2007)
7. Beaudouin-Lafon, M.: Instrumental interaction: An interaction model for designing post-WIMP user interfaces. In: Proceedings of the SIGCHI Conference on Human Factors in Computing Systems, CHI '00, pp. 446–453. ACM, New York, NY, USA (2000)
8. Benade, A.H.: Fundamentals of Musical Acoustics. Dover, New York (1976)
9. Benson, D.: Music, A Mathematical Offering. Cambridge University Press, Cambridge (2006)
10. Bogert, B., Healy, M., Tukey, J.: The quefrency alanysis of time series for echoes: cepstrum, pseudo-autocovariance, cross-cepstrum and saphe cracking. In: Proceedings Symposium on Time Series Analysis, pp. 209–243 (1963)
11. Bracewell, R.: The Fourier Transform and Its Applications. Electrical Engineering Series. McGraw-Hill, New York (2000)
12. Brandtsegg, Ø., Saue, S., Johansen, T.: Particle synthesis, a unified model for granular synthesis. In: Proceedings of the Linux Audio Conference 2011 (2011)
13. Brun, M.L.: A derivation of the spectrum of FM with a complex modulating wave. Computer Music Journal $1(4)$, 51–52 (1977)
14. Cherniakov, M.: An Introduction to Parametric Digital Filters and Oscillators. John Wiley & Sons, New York (2003)
15. Chowning, J.: The synthesis of complex spectra by means of frequency modulation. Journal of the Audio Engineering Society $21(7)$, 526–534 (1973)
16. Chu, E., George, A.: Inside the FFT Black Box: Serial and Parallel Fast Fourier Transform Algorithms. Computational Mathematics. Taylor & Francis (1999)
17. Cook, P.: A toolkit of audio synthesis classes and instruments in C++ (1995). URL https://ccrma.stanford.edu/software/stk/
18. Cooley, J.W., Tukey, J.W.: An algorithm for the machine calculation of complex Fourier series. Mathematics of Computation $19(90)$, 297–301 (1965)
19. Damasevicius, R., Stuikys, V.: Separation of concerns in multi-language specifications. Informatica $13(3)$, 255–274 (2002)

© Springer International Publishing AG 2017
V. Lazzarini, *Computer Music Instruments*,
https://doi.org/10.1007/978-3-319-63504-0

20. Dattorro, J.: Effect design, part 2: Delay line modulation and chorus. Journal of the Audio Engineering Society **45**(10), 764–788 (1997)
21. Dirichlet, P.G.L.: Sur la convergence des séries trigonometriques qui servent à répresenter une fonction arbitraire entre des limites donneés. Journal für die reine und angewandte Mathematik **4**, 157–169 (1829)
22. Dodge, C., Jerse, T.A.: Computer Music: Synthesis, Composition and Performance, 2nd edn. Schirmer, New York (1997)
23. Dolson, M.: The phase vocoder: A tutorial. Computer Music Journal **10**(4), 14–27 (1986)
24. Dougherty, D.: The Maker movement. Innovations **7**(3), 11–14
25. Flanagan, F., Golden, R.: Phase vocoder. Bell System Technical Journal **45**, 1493–1509 (1966)
26. Fourier, J.B.: Théorie analytique de la chaleur. Chez Firmin Didot, Père et fils, Paris (1822)
27. Gardner, W.G.: Efficient convolution without input-output delay. Journal of the Audio Engineering Society **43**(3), 127–136 (1995)
28. Gentner, D., Nielsen, J.: The Anti-Mac interface. Commun. ACM **39**(8), 70–82 (1996)
29. Grey, J.: An Exploration of Musical Timbre Using Computer-based Techniques. Department of Psychology, Stanford University, Stanford (1975)
30. Harris, F.: On the use of windows for harmonic analysis with the discrete Fourier transform. Proceedings of the IEEE **66**(1), 51–83 (1978)
31. Howard, D., Angus, J.: Acoustics and Psychoacoustics. Focal Press (2009)
32. IEEE: IEEE standard for floating-point arithmetic. IEEE Std 754-2008 pp. 1–70 (2008)
33. Jaffe, D.A.: Spectrum analysis tutorial, part 1: The discrete Fourier transform. Computer Music Journal **11**(2), 9–24 (1987)
34. Jaffe, D.A.: Spectrum analysis tutorial, part 2: Properties and applications of the discrete Fourier transform. Computer Music Journal **11**(3), 17–35 (1987)
35. Jayant, N.S., Noll, P.: Digital Coding of Waveforms: Principles and Applications to Speech and Video. Prentice Hall Professional Technical Reference, Englewood Cliffs, NJ (1990)
36. Keller, D., Lazzarini, V., Pimenta, M.S. (eds.): Ubiquitous Music. Computational Music Science. Springer (2014)
37. Kleimola, J., Lazzarini, V., Timoney, J., Valimaki, V.: Vector phase shaping synthesis. In: Proceedings of the 14th International Conference on Digital Audio Effects (DAFx-11), pp. 223–240. Paris, France (2011)
38. Kleimola, J., Lazzarini, V., Vämäki, V., Timoney, J.: Feedback amplitude modulation synthesis. EURASIP Journal Adv. Sig. Proceedings **2011**:434378 (2011)
39. Kushner, D.: The making of Arduino. IEEE Spectrum URL http://spectrum.ieee.org/geek-life/hands-on/the-making-of-arduino/
40. Lazzarini, V.: Introduction to digital audio signals. In: R. Boulanger, V. Lazzarini (eds.) The Audio Programming Book, pp. 431–462. MIT Press, Cambridge, MA (2010)
41. Lazzarini, V.: Programming the phase vocoder. In: R. Boulanger, V. Lazzarini (eds.) The Audio Programming Book, pp. 557–580. MIT Press, Cambridge, MA (2010)
42. Lazzarini, V.: Spectral audio programming basics: The DFT, the FFT, and convolution. In: R. Boulanger, V. Lazzarini (eds.) The Audio Programming Book, pp. 521–538. MIT Press, Cambridge, MA (2010)
43. Lazzarini, V.: The STFT and spectral processing. In: R. Boulanger, V. Lazzarini (eds.) The Audio Programming Book, pp. 539–556. MIT Press, Cambridge, MA (2010)
44. Lazzarini, V.: Time-domain audio programming. In: R. Boulanger, V. Lazzarini (eds.) The Audio Programming Book, pp. 463–520. MIT Press, Cambridge, MA (2010)
45. Lazzarini, V.: The development of computer music programming systems. Journal of New Music Research **42**(1), 97–110 (2013)
46. Lazzarini, V., Costello, E., Yi, S., ffitch, J.: Development tools for ubiquitous music on the World Wide Web. In: Ubiquitous Music, pp. 111–128. Springer (2014)
47. Lazzarini, V., J., T., Pekonen, J., Valimaki, V.: Adaptive phase distortion synthesis. In: Proceedings of the 12th International Conference on Digital Audio Effects, pp. 28–35. Milan Institute of Technology, Como, Italy (2009)

48. Lazzarini, V., Keller, D., Pimenta, M.S., Timoney, J.: Ubiquitous music ecosystems: Faust programs in Csound. In: Ubiquitous Music, pp. 129–150. Springer (2014)

49. Lazzarini, V., Kleimola, J., Timoney, J., Valimaki, V.: Five variations on a feedback theme. In: Proceedings of the 12th International Conference on Digital Audio Effects, pp. 139–145. Milan Institute of Technology, Como, Italy (2009)

50. Lazzarini, V., Kleimola, J., Timoney, J., Valimaki, V.: Aspects of second-order feedback AM synthesis. In: Proceedings of the International Computer Music Conference, pp. 92–98 (2011)

51. Lazzarini, V., Timoney, J.: New methods of formant analysis-synthesis for musical applications. In: Proceedings International. Computer Music Conference, pp. 239–242. Montreal, Canada (2009)

52. Lazzarini, V., Timoney, J.: New perspectives on distortion synthesis for virtual analog oscillators. Computer Music Journal 34(1), 28–40 (2010)

53. Lazzarini, V., Timoney, J.: Theory and practice of modified frequency modulation synthesis. Journal of the Audio Engineering Society 58(6), 459–471 (2010)

54. Lazzarini, V., Timoney, J.: Synthesis of resonance by nonlinear distortion methods. Computer Music Journal 37(1), 35–43 (2013)

55. Lazzarini, V., Timoney, J., Lysaght, T.: Spectral processing in Csound 5. In: Proceedings of International Computer Music Conference, pp. 102–105. New Orleans, USA (2006)

56. Lazzarini, V., Timoney, J., Lysaght, T.: Asymmetric-spectra methods for adaptive FM synthesis. In: Proceedings of the 11th International Conference on Digital Audio Effects (DAFx-11), pp. 42–49. Espoo, Finland (2008)

57. Lazzarini, V., Timoney, J., Lysaght, T.: The generation of natural-synthetic spectra by means of adaptive frequency modulation. Computer Music Journal 32(2), 9–22 (2008)

58. Lazzarini, V., Timoney, J., Lysaght, T.: Nonlinear distortion synthesis using the split-sideband method, with applications to adaptive signal processing. Journal of the Audio Engineering Society 56(9), 684–695 (2008)

59. Lazzarini, V., Yi, S., ffitch, J., Heintz, J., Brandtsegg, Ø., McCurdy, I.: Csound: A Sound and Music Computing System. Springer, Berlin (2016)

60. Lazzarini, V., Yi, S., Timoney, J.: Web audio: Some critical considerations. In: Sixth Workshop on Ubiquitous Music, pp. 1–12. Linnaeus University, Växjö, Sweden (2015)

61. Le Brun, M.: Digital waveshaping synthesis. Journal of the Audio Engineering Society 27(4), 250–266 (1979)

62. Loy, G.: The CARL system: Premises, history and fate. Computer Music Journal 26(4), 23–54 (2002)

63. Mathews, M.: An acoustical compiler for music and psychological stimuli. Bell System Technical Journal 40(3), 553–557 (1961)

64. Mathews, M., Miller, J.E.: MUSIC IV Programmer's Manual. Bell Telephone Lab, Murray Hill, N.J. (1964)

65. Mathews, M., Miller, J.E., Moore, F.R., Pierce, J.R.: The Technology of Computer Music. MIT Press, Cambridge, MA (1969)

66. McAulay, R., Quatieri, T.: Speech analysis/synthesis based on a sinusoidal representation. IEEE Transactions on Acoustics, Speech and Signal Processing 34(4), 744–754 (1986)

67. MIDI Manufacturers Association: Midi 1.0 specification (1983). URL http://www.midi.org

68. Moore, F.R.: Elements of Computer Music. Prentice-Hall, Inc., Upper Saddle River, NJ, USA (1990)

69. Moorer, J.: Signal processing aspects of computer music: A survey. Proceedings of the IEEE 65(8), 1108–1137 (1977)

70. Moorer, J.A.: The synthesis of complex audio spectra by means of discrete summation formulas. Journal of the Audio Engineering Soc 24(9), 717–727 (1976)

71. Moorer, J.A.: The use of the phase vocoder in computer music applications. Journal of the Audio Engineering Society 26(1/2), 42–45 (1978)

72. Mulgrew, B., Grant, P., Thompson, J.: Digital Signal Processing: Concepts and Applications. Macmillan Press, London (1999)

73. Nielsen, J.: Noncommand user interfaces. Communications ACM **36**(4), 83–99 (1993)
74. Noble, J., Joshua, N.: Programming Interactivity: A Designer's Guide to Processing, Arduino, and Openframeworks, 1st edn. O'Reilly Media, Inc. (2009)
75. Nyquist, H.: Certain topics in telegraph transmission theory. Transactions of the AIEE **47**, 617–644 (1928)
76. Oppenheim, A.V., Schafer, R.W., Buck, J.R.: Discrete-time Signal Processing (2nd Ed.). Prentice-Hall, Inc., Upper Saddle River, NJ, USA (1999)
77. Orlarey, Y., Fober, D., Letz., S.: Faust: An efficient functional approach to DSP programming. In: G. Assayag, A. Gerszo (eds.) New Computational Paradigms for Computer Music. Edition Delatour (2009)
78. Ousterhout, J.: Scripting: higher-level programming for the 21st century. IEEE Computer **31**(3), 23–30 (1998)
79. Palamin, J.P., Palamin, P., Ronveaux, A.: A method of generating and controlling musical asymmetrical spectra. Journal of the Audio Engineering Society **36**(9), 671–685 (1988)
80. Pampin, J.: ATS: A system for sound analysis transformation and synthesis based on a sinusoidal plus crtitical-band noise model and psychoacoustics. In: Proceedings of the International Computer Music Conference, pp. 402–405. Miami, FL (2004)
81. Parallax Inc.: The original BASIC stamp microcontroller (2016). URL https://www.parallax.com/microcontrollers/basic-stamp
82. Phillips, D.: Linux Music & Sound. No Starch Press, San Francisco, CA (2000)
83. Poepel, C., Dannenberg, R.: Audio signal driven sound synthesis. In: Proceedings of the 2005 International Computer Music Conference, pp. 391–394. Barcelona, Spain
84. Puckette, M.: Formant-based audio synthesis using nonlinear distortion. Journal of the Audio Engineering Society **43**(1/2), 40–47 (1995)
85. Puckette, M.: Phase-locked vocoder. In: IEEE ASSP Workshop onApplications of Signal Processing to Audio and Acoustics, pp. 222–225 (1995)
86. Risset, J.C.: An Introductory Catalogue of Computer Synthesized Sounds. Bell Telephone Lab, Murray Hill, N.J. (1969)
87. Roads, C.: Microsound. MIT Press, Cambridge, MA (2001)
88. Roads, C., Mathews, M.: Interview with Tongues. Computer Music Journal **4**(4), pp. 15–22 (1980)
89. Rocher, M.: Introduction to the theory of Fourier's series. Annals of Mathematics **7**(3), 81–152 (1906)
90. Rodet, X.: Time domain formant-wave-function synthesis. Computer Music Journal **8**(3), 9–14 (1984)
91. Roebel, A.: Efficient spectral envelope estimation and its application to pitch shifting and envelope preservation. In: Proceedings of the 8th International Conference on Digital Audio Effects, pp. 30–35 (2005)
92. Schottstaedt, B.: The simulation of natural instrument tones using frequency modulation with a complex modulating wave. Computer Music Journal **1**(4), 46–50
93. Serra, X., Smith, J.: Spectral modeling synthesis: A sound analysis/synthesis based on a deterministic plus stochastic decomposition. Computer Music Journal **14**, 12–24 (1990)
94. Shannon, C.E.: Communication in the presence of noise. Proceedings Institute of Radio Engineers **37**(1), 10–21 (1949)
95. Shneiderman, B.: Designing the User Interface: Strategies for Effective Human-computer Interaction. Addison-Wesley Longman Publishing Co., Inc., Boston, MA, USA (1986)
96. Smalley, D.: Spectro-morphology and structuring processes. In: S. Emmerson (ed.) The Language of Electroacoustic Music, pp. 61–96. Macmillan Press, London (1986)
97. Smith, J., Serra, X.: PARSHL: An analysis/synthesis program for non-harmonic sounds based on a sinusoidal representation. In: International Computer Music Conference, pp. 290–297. Urbana, Illinois, USA (1987)
98. Steiglitz, K.: A Digital Signal Processing Primer, with Applications to Digital Audio and Computer Music. Addison-Wesley Longman, Redwood City (1996)

99. Stockham Jr., T.G.: High-speed convolution and correlation. In: Proceedings of the April 26-28, 1966, Spring Joint Computer Conference, AFIPS '66, pp. 229–233. ACM, New York (1966)
100. Tenney, J.: Sound generation by means of a digital computer. Journal of Music Theory **7**(1), 24–70 (1963)
101. Timoney, J., Pekonen, J., Lazzarini, V., Välimäki, V.: Dynamic signal phase distortion using coefficient-modulated allpass filters. Journal of the Audio Engineering Society **62**(9), 596–610 (2014)
102. Tomisawa, N.: Tone production method for an electronic musical instrument (1981). US Patent 4,249,447
103. Torger, A., Farina, A.: Real-time partitioned convolution for ambiophonics surround sound. In: 2001 IEEE Workshop on Applications of Signal Processing to Audio and Acoustics. New Paltz, New York (2001)
104. Widrow, B., Kollár, I.: Quantization Noise: Roundoff Error in Digital Computation, Signal Processing, Control, and Communications. Cambridge University Press, Cambridge, UK (2008)
105. Windham, G., Steiglitz, K.: Input generators for digital sound synthesis. Journal of the Acoustic Society of America **47**(2), 665–6
106. Wishart, T.: Audible Design. Orpheus the Pantomime, York (1994)
107. Wood, A.: The Physics of Music. Chapman and Hall, London (1975)
108. Wright, M., Freed, A.: Open Sound Control: A new protocol for communicating with sound synthesizers. In: Proceedings of the ICMC, pp. 101–104. Thessaloniki, Greece (1997)
109. Wright, M., Freed, A., Momeni, A.: Open Sound Control, state of the art 2003. In: Proceedings of the 2003 Conference on New Interfaces for Musical Expression (NIME-03), pp. 153–159. Montreal, Canada (2003)

Index

© Springer International Publishing AG 2017
V. Lazzarini, *Computer Music Instruments*,
https://doi.org/10.1007/978-3-319-63504-0

Printed in the United States
By Bookmasters